M000306839

MATH GIRLS³

DATE DUE

INCOM ... IS

∀

HI

TRANS

MATH GIRLS 3: GÖDEL'S INCOMPLETENESS THEOREMS
by Hiroshi Yuki

Originally published as *Sūgaku Gāru Gēderu No Fukanzensei Teiri*
Copyright © 2009 Hiroshi Yuki
Softbank Creative Corp., Tokyo

English translation © 2016 by Tony Gonzalez
Edited by Joseph Reeder and M.D. Hendon
Additional editing by Michael Klipper
Cover design by Kasia Bytnerowicz

Published 2016 by

Bento Books, Inc.
Austin, Texas 78732

bentobooks.com

ISBN 978-1-939326-27-0 (hardcover)
ISBN 978-1-939326-28-7 (trade paperback)
ISBN 978-1-939326-29-4 (case laminate)
Library of Congress Control Number: 2014958330

Printed in the United States of America
First edition, May 2016

3 0005 00015084 0

Math Girls[3]:
Gödel's Incompleteness Theorems

Contents

To my readers

This book contains math problems covering a wide range of difficulty. Some will be approachable by elementary school students, while others are so deep they have shaken the very foundations of mathematics.

The characters often use words and diagrams to express their thoughts, but in some places equations tell the tale. If you find yourself faced with math you don't understand, feel free to skip over it and continue on with the story. Tetra and Yuri will be there to keep you company.

If you have some skill at mathematics, then please follow not only the story, but also the math. You might be surprised at what you discover. You may even stumble upon secrets yet to be imagined.

—Hiroshi Yuki

Prologue

The waves roll in, then back out. Repetition upon repetition.

The rhythm of this repetition turns consciousness
to the self.

The rhythm of this repetition turns consciousness
to the past.

I recall those days when we had wings to soar the sky. Yet I remained
a tiny bird fluttering in a cage.

A time of which I long to speak.

A time of which I must remain silent.

When spring arrives, I think of math.

Symbols on paper, describing space.

Equations on paper, deriving truths.

When spring arrives, I think of them.

Talking about math in the library.

Afternoons spent beneath the warm spring sun.

Let me tell you a tale, about how they taught me to spread my wings and fly.

CHAPTER 1

Liar, Liar

"Mirror, mirror on the wall, who's the fairest one of all?"
"Thou, O Queen, art the fairest of all."
The Queen was contented, because she knew the mirror could speak nothing but the truth.

GRIMM'S FAIRY TALES
Snow White, trans. by L.L. Weedon

1.1 WHO TELLS THE TRUTH?

1.1.1 *Mirror, Mirror*

"You've read Snow White, right?" Yuri asked.

"Sure," I said. "That's the one about the pumpkin-riding princess with the fragile footwear."

"That's Cinderella, doofus."

"You sure about that?"

I tried to keep a straight face, but Yuri saw through me and punched my arm with a laugh.

It was early January, near the end of our winter break. There would be exams waiting when school started back up, but I wasn't in the mood to study.

I was in the eleventh grade; Yuri was in the eighth. She was my cousin on my mother's side, but she hung around my house often enough that people mistook her for my sister. Her house was just down the street, so we'd been playmates for as long as either of us could remember. On days when we didn't have school, she'd taken to lounging in my room reading my books while I studied. Or tried to study, at least.

"Snow White had the evil stepmother that talked to mirrors. You know, 'Mirror, mirror on the wall, who's the fairest one of all?'"

"Ah, right. The one with the beauty detector."

"I don't know if I'd have the guts to ask a question like that."

Yuri pulled her chestnut brown ponytail around for a closer look. She sighed.

"I wish Mom would let me dye my hair, at least. And wear makeup."

I took an appraising glance at Yuri, unsure why she'd be so self-conscious about her looks. She seemed cute enough to me. Her expression was always changing, which was fun to watch. She had a personality that made me think of popcorn bursting. But most important of all she was smart and interesting to talk to.

"I don't think you have anything to worry about," I said. "You're fine just the way you are."

Yuri blinked.

"A compliment? From *you*?" She squinted and leaned toward me. "Who are you, and what have you done with my cousin?"

"Hey, kids!" my mother shouted from the kitchen. "Who wants bagels?"

Yuri's expression flipped to one of delight.

"I do!" she shouted.

She stood and yanked me toward the door. The promise of food always brought out the strength in her.

"C'mon! Let's eat!"

1.1.2 Finding Truth Tellers

In the dining room, Yuri picked up a book of logic problems I'd left lying on the table.

"This any good?" she asked.

"Dunno. I borrowed it from the school library before vacation, but I never got around to reading it."

"Huh. Your school's got much cooler books than mine does."
Yuri flipped through the pages.
"Here's a good one," she said.

Who's telling the truth?
A_1 says, "One of us is a liar." A_2 says, "Two of us are liars." A_3 says, "Three of us are liars." A_4 says, "Four of us are liars." A_5 says, "Five of us are liars." Who's telling the truth?

My mother came in from the kitchen, carrying a platter of bagels.
"Take whichever you like," she said, holding it out. "I've got plain, walnut, basil . . ."
"What's this one?" Yuri asked.
"Onion."
"Dibs!" Yuri said, grabbing it.
"What about you?" Mom asked, turning to me. The bagels smelled warm and fresh.
"Whatever's fine. So Yuri, about that problem—"
"Pick one," Mom insisted, thrusting the plate my way.
"Uh, plain then." I started reaching for a bagel. "So anyway, Yuri—"
"You should try the walnut," Mom said.
"But you just said— Fine, fine."
I took the walnut bagel and my mother retreated to the kitchen, a satisfied look on her face.
"So *anyway*," I finally managed, "this is one of those problems with liars and truth-tellers, right? Truth-tellers always tell the truth—"
"—and liars gonna lie, right. And every one of these five people A_1 through A_5 is either a liar or a truth-teller."
"That's easy, then," I said. "A_4 is telling the truth, and the rest are lying."

"Well that's no fun, if you know the answer right off."

"You just have to think in cases," I said. "There has to be between 0 and 5 people telling the truth. It's not possible that nobody is telling the truth, because that would make A_5's statement true. Then A_5 is a truth-teller, contradicting the assumption that there were no truth-tellers."

Yuri nodded. "Okay, I follow that."

"So what if there was one truth-teller? That would mean there's four liars, which makes A_4's statement true. One truth-teller, four liars—no contradiction."

"Good for you, A_4," Yuri said.

"Next is the case where there are two truth-tellers and three liars. Then A_3's statement should be true, but A_3 is the only one saying it so there would have to be four liars, another contradiction. The same thing happens with 3, 4, and 5 truth-tellers, so the only possible answer is A_4."

I paused and looked at the problem again.

"Hmm..."

"'Hmm' what?" Yuri asked.

"I just noticed that they used subscripted names like A_1 and A_2 and so on, instead of just A, B, C ..."

"So?"

"I think they were leaving it open for generalization. Do you follow this?"

I grabbed a pen and started scribbling on a napkin.

Who's telling the truth? (Generalized)

B_1 says, "1 of us is a liar."
B_2 says, "2 of us are liars."
B_3 says, "3 of us are liars."

$$\vdots$$

B_{n-1} says, "$(n-1)$ of us are liars."
B_n says, "n of us are liars."

Who's telling the truth?

"What's with the n's?" Yuri asked around a mouthful of bagel.

"That's a good thing to ask. In this case, it stands for any natural number."

"Well that's not helpful. So you can have, like, an infinity of liars?"

"No, because infinity isn't a number. You can make n really, really big, but there will still just be n liars or truth-tellers, a finite number."

"Darn."

"The problem's still just as easy, though. You can solve it the same way as before."

"Yeah? Oh, so B_{n-1} would be the truth-teller?"

"Smart girl."

"Smart would be telling you there will always be 1 truth-teller and $(n-1)$ liars. Oops, there I go again, showing off my brains."

Yuri gave her ponytail a flip.

"Double smart would be calling this 'generalization through the introduction of a variable,'" I said.

"That's the n, yeah? And n can be any number I want?"

"Well, like I said, it has to be a natural number. But yeah, any integer 1 or larger."

"Well then I'm *triple* smart, because I noticed that n can't be 1, and you didn't."

Who's telling the truth? (with $n = 1$)

C_1 says, "1 of us is a liar."

"How so?" I asked. "That's just the case where there are no truth-tellers."

"So what's C_1 then?"

"Well he'd have to be a liar."

"Which means he's telling the truth, so he has to be a truth-teller!"

"Which ... would be a contradiction, so the problem has no solution. Huh."

"There can be a problem with no solution?"

"A poorly posed one, sure. Here, the condition that everyone has to be either a liar or a truth-teller is messing things up. The problem is unsolvable when $n = 1$, because we can't classify C_1 as either one. Nice catch, Yuri. You've beat me again."

"That feels good, but even better would be to find an answer. I don't like leaving things unresolved like that."

"I don't, either."

"Hey, I know! There *is* a liar here! It's whoever wrote the problem this way!"

1.1.3 Same Answers

"Hey," I said. "I just thought of a better logic problem. Try this one."

A question with identical answers

Think of a yes–no question that both a liar and a truth-teller would answer in the same way.

"I don't get it," Yuri said. "What do you mean, answer the same way?"

"Just that. If a liar answered the question 'yes,' then a truth-teller would have to say 'yes,' too. And if one answered 'no' then so would the other."

"You sure this problem has an answer?"

I nodded.

"Okay, gimme a second."

Yuri scrunched up her face in concentration. After a short time she brightened back up.

"Oh, that's not so hard. You just have to ask 'Do you tell the truth?' They'd both have to answer 'yes.'"

"Well done. The opposite works, too. If you ask 'Do you tell lies?' they'd both answer 'no.'"

My mother came back into the room.

"Who wants some hot chocolate?"

"I'd rather have coffee," I said, "but whatever."

"I love your hot chocolate!" Yuri said.

"Such a sweet child," my mother replied.

Yuri turned back to me. "These are really weird problems if you think about it. Could you imagine being a truth-teller? Like, every time you open your mouth, you can only say true things? That would be so weird."

"It's also weird that the liars have to have the same superpower as the truth-tellers," I said.

"What superpower?"

"Omniscience. If you want to be sure you always lie, you have to know everything so that you don't slip up and tell the truth."

"Slip up and tell the truth!" Yuri laughed. "I like that."

My mother frowned.

"I hope I have two truth-tellers in front of me."

1.1.4   Answering with Silence

"Hey, I've got one!" Yuri said. "In the last one liars and truth-tellers both had to give the same answer, right? Well can you think of a question like this?"

An unanswerable question

Think of a yes–no question that a liar can answer, but a truth-teller cannot.

"Hmm, tricky one," I said, scratching my jaw. "Maybe ask them a question that nobody knows the answer to? Something like 'Are there infinitely many twin primes?'"

"What's a twin prime?"

"A pair of prime numbers that are two apart, like 3 and 5, or 5 and 7. Nobody on earth knows if there are infinitely many of them."

Yuri shook her head. "Nah, not good enough. Even if nobody knows now, someday they might. Besides, even now the liar couldn't answer that one."

"Yeah, I guess you're right," I admitted. "Hmm . . . "

"Here's the question I was thinking of: 'Will you answer this question 'No'?"

"Ooh, good one! If a truth-teller answered 'yes' then he'd be lying because he didn't say 'no,' and if he answered 'no' he'd be lying because he didn't say 'yes.' Wait, did I say that right? I'm starting to confuse myself."

"Yeah, I think so... Anyway the point is that the truth-teller can't give a true answer, but the liar doesn't have a problem—he can just say 'yes,' because that's a lie."

"Or 'no'! That's a lie, too!"

"Hey, yeah. Wow, this problem hurts my brain."

"Mine, too."

1.2   Logic Problems

1.2.1   Alice, Boris, and Chris

"These are fun!" Yuri said, flipping through the book. "Let's do some more!"

She flipped through the book and stopped on a page, grinning. "How about this one?"

Three people's belongings

Alice, Boris, and Chris each have a hat, a watch, and a jacket, each of which is red, green, or yellow. No two of the same kind of object are the same color, and nobody has two items that are the same color.

Given the following, figure out what color each person's items are:

· Alice's watch is yellow.

· Boris's watch is not green.

· Chris's hat is yellow.

"What's so funny about this problem?" I asked.

Yuri rolled her eyes.

"Their fashion sense, of course! I mean, just imagine what these three must look like!"

"Colorful would be my guess. But can you solve the problem?"

"Probably. Too much work, though. I'll pass."

"It's not that bad. You just have to make a table."

1.2.2 Thinking with Tables

"So we need to make a table of what everyone has," I said. "Start with the given conditions."

	Hat	Watch	Jacket
Alice		yellow	
Boris		not green	
Chris	yellow		

Given conditions

Yuri put on her glasses and peered at the table.

"By 'given conditions,' you mean what the problem already said, right?"

"Right. When you have to deal with a lot of information, it's way easier to lay it out in a table than juggle it all in your head. See how we can get the color of Boris's watch right away? It isn't green, and it has to be a different color than Alice's, so it isn't yellow. That just leaves red."

"Makes sense."

	Hat	Watch	Jacket
Alice		yellow	
Boris		red	
Chris	yellow		

Filling in Boris's watch

"Now you can figure out what color Boris's jacket must be. Do you see it?"

"Yellow, right?" Yuri said.

"Right. How come?"

"Because it has to be yellow."

"Yeah, but *why* does it have to be yellow?"

"Because the other yellows are taken. Alice's watch is yellow, and so is Chris's hat. That means neither Alice or Chris can have a yellow jacket. Only Boris can."

"Now *that's* a good explanation.

	Hat	Watch	Jacket
Alice		yellow	Can't be yellow
Boris		red	Must be yellow
Chris	yellow		Can't be yellow

Neither Alice nor Chris can have a yellow jacket

	Hat	Watch	Jacket
Alice		yellow	
Boris		red	**yellow**
Chris	yellow		

Filling in Boris's jacket

"Now you can get Chris's watch." I said.

"Yep. Alice's watch is yellow, and Boris's is red, so Chris's must be green."

"Exactly."

	Hat	Watch	Jacket
Alice		yellow	
Boris		red	yellow
Chris	yellow	**green**	

Filling in Chris's watch

"Check it out! Now we can get Chris's jacket." Yuri pointed at Chris's row in the table. "It can't be yellow, because his hat is, and it can't be green, because his watch is. So his jacket must be red."

	Hat	Watch	Jacket
Alice		yellow	
Boris		red	yellow
Chris	yellow	green	**red**

Filling in Chris's jacket

"Hmm, what can we do now..." I said, scanning the columns and rows.

"Alice's hat! See? Boris's watch is red, and so is Chris's jacket. That means we can't use red for Alice's watch or jacket, so we have to use it for her hat!"

	Hat	Watch	Jacket
Alice	**red**	yellow	
Boris		red	yellow
Chris	yellow	green	red

Filling in Alice's hat

"Nice!" I said. "Let's see, so next—"
"Quiet, I want to do the rest! I think I got Alice's jacket..."

	Hat	Watch	Jacket
Alice	red	yellow	**green**
Boris		red	yellow
Chris	yellow	green	red

Filling in Alice's jacket

"And last is Boris's hat."

	Hat	Watch	Jacket
Alice	red	yellow	green
Boris	**green**	red	yellow
Chris	yellow	green	red

Filling in Boris's hat

"Done!"

"Good job!" I said.

1.2.3   Well-designed Problems

"Meh, that was too easy," Yuri said.

"Sure looked like you were having fun, though," I said. "I kinda like this problem. It's well designed."

"In what way?"

"The conditions you're given at the beginning, mainly."

I pointed at the page.

· Alice's watch is yellow.

· Boris's watch is not green.

· Chris's hat is yellow.

"There's not too many, not too few."

"I have no idea what you're talking about."

"I mean that if there were more than this, the problem would be trivial. Take one away, and it's unsolvable."

Yuri frowned. "You sure about that?"

She pulled the napkin closer and picked up the pen. After a minute of scribbling she put it back down with a grin.

"You don't need all this. Try taking away the condition that Chris's hat is yellow."

· Alice's watch is yellow.

· Boris's watch is not green.

"You can solve the problem two ways if you do that," Yuri said.

	Hat	Watch	Jacket			Hat	Watch	Jacket
Alice	red	yellow	green		Alice	green	yellow	red
Boris	green	red	yellow		Boris	yellow	red	green
Chris	yellow	green	red		Chris	red	green	yellow

Answers without the condition on Chris's hat

"Ah, okay. Sorry, I didn't phrase that well. I should have said you can't get a *unique* solution."

"Why should it be unique?"

"Well, if I were the one making the problem, I'd want to challenge the person solving it to find the answer I was thinking of. Solving problems is fun, but creating them can be just as interesting. So when I find a good problem, I wonder how I might make a similar one."

1.3 WHAT COLOR IS YOUR HAT?

1.3.1 *What Others Don't Know*

"Wow, here's a weird one," Yuri said, showing me a page in the book of logic problems.

What color is the hat?

Amy, Bob, and Clara are on a game show. The host sits them down and explains the game:

Host We'll be putting one of five hats on each of you. Three of the hats are red, and two are white. You won't be able to see the color of the hat you're wearing, but you can see the others.

The Host puts a hat on each contestant and hides the remaining two.

Host "Amy, what color is your hat?"

Amy "I don't know."

Host "Bob, what color is your hat?"

Bob "I don't know."

Clara can see that both Amy and Bob are wearing red hats. The MC turns to her.

Host "Clara, what color is your hat?"

Can Clara give a guaranteed correct answer?

"What a strange game," Yuri said.

I nodded. "Indeed. Lemme give it a shot, though. Let's see..."

Clara can see that both Amy and Bob are wearing red hats. There are three red hats, so her hat could be either red or white. The only clue is that neither Amy nor Bob know what color hat they're wearing. Hmm...

"Got it?" Yuri asked.

"Still thinking."

Amy can see Bob and Clara. Amy doesn't know what color hat she's wearing, so Bob and Clara can't both be wearing white hats—otherwise she'd know her hat was red. That means at least one of Bob or Clara is wearing a red hat.

Bob can see Amy and Clara, so by a similar line of thinking at least one of Amy and Clara have to be wearing a red hat. So... so what does that tell me? Wow, this problem is harder than it looked...

"You're *still* thinking?" Yuri said, smirking.

"What, you've already figured it out?"

"It's not all that hard."

Looks like I've got some catching up to do. Time to get organized, examine things in cases. Let's see... Clara's hat has to be either white or red, so...

Assume Clara's hat is white—

· Amy looks at Bob (red) and Clara (white). She can't tell what color her own hat is.

· Bob looks at Amy (red) and Clara (white)...

Hang on, now. Clara isn't the only one getting clues. When Bob hears Amy's response, he thinks like this...

If Clara's hat is white, Bob thinks—

· Amy looked at me (unknown color) and Clara (white).

· Amy said she doesn't know what color hat she's wearing.

· That means at least one of Clara or me must have a red hat.

· Clara's hat is white, so mine must be red!

So in this case, Bob would have known he's wearing a red hat... But that's not what happened!

· Bob said "I don't know".

· That means the assumption that Clara's hat is white must be wrong.

· Clara's hat must therefore be red!

"Got it," I said. "Clara's hat is red."

"It's about time, slowpoke."

1.3.2   A Problem for the Problem Writer

When I explained my train of thought to Yuri, she frowned.

"Hang on, maybe you missed something," she said.

"What's that?"

"Shouldn't you also walk through the case where you assume Clara's hat is red?"

"Good point," I said. "But in this case there's no need. The problem says that the hats have to be red or white, so I've covered all the bases just by showing that it isn't white."

"But what if whoever wrote this problem was a liar, instead of a truth-teller? You're assuming that Clara has to have a red or a white hat, and that since her hat isn't white it must be red. But what if the condition that all hats are red or white was a lie?"

"Interesting. I don't think you have to worry about problems like this intentionally lying to you, but I suppose there's always the chance that the problem writer goofed. Or *I* did. Maybe it's worth walking through the 'Clara's hat is red' case, just to make sure."

"Let's do it!"

Assume Clara's hat is red—

- Amy looks at Bob (red) and Clara (red). She can't tell what color her own hat is.

  → Amy says "I don't know."

- Bob looks at Amy (red) and Clara (red). He can't tell what color his own hat is.

  → Bob says "I don't know."

"Looks like everything works out," I said. "There aren't any contradictions like when we assumed Clara's hat was white, since Bob doesn't have enough information to give an answer. So it seems there's no problem with the condition that Clara's hat must be either red or white."

"Well that's a relief. Tricky problem, but that was fun."

"It was. I like that 'I don't know' ended up being a big clue. Also that you have to step into everybody's shoes and look at things from their perspective."

"Sure beats solving for x."

"By the way, how'd you get the answer so fast?"

"Umm..." Yuri pursed her lips as she thought. "It's hard to explain. I could work it out in my head, but I'm not good at the step-by-step thing like you. Saying mathy stuff like 'this and that implies one or the other' just doesn't come natural to me."

"You'll get there. Hey, I just realized there's a neat theorem in this hat-problem world: when one person says 'I don't know,' at least one of the other two is wearing a red hat."

"See? Exactly the kind of thing I'm talking about."

"Let's head back up to my room." I glanced back at the kitchen. "Thanks for the bagels, mom!"

"Yeah, they were delicious!"

"I'll be up with tea in a bit!" Mom shouted back.

1.3.3 Reflections

As we walked back into my room, Yuri snapped her fingers.

"I just thought of an easier way to solve the hat problem."

"How?" I asked.

"Hang mirrors on the wall!"

"That's a way to cheat the problem, not solve it."

"Speaking of which, why don't you have any mirrors in your room? Are you a vampire or something?"

"Vampires don't hang mirrors in their room?"

"Because they don't cast reflections. Don't you know anything?"

"Light reflects off of them, but the light they reflect won't reflect off a mirror? That doesn't make any sense."

"Bah, never mind. I forgot you're culturally challenged. Besides, I've got my own."

Yuri pulled a small mirror out of her bag.

"You carry a mirror with you?"

"Well duh," Yuri said, rolling her eyes.

Yuri silently examined her face and hair from various angles.

"Checking to see if you're becoming a vampire?"

"Busy here. Leave me alone."

I sighed and turned to my desk.

"Hey, I know how to beat the talking mirror," she said. "You just have to live in a world with no other people, so nobody can be fairer than you. But I guess there'd be no point if there was no one around to see you..."

Yuri stood in the middle of the room, holding up her mirror in a dramatic pose.

"Mirror, mirror, in my hand, who's the fairest in the land?"

My mother entered the room with a tray of hot tea.

"Oh, how cute," she said. "You're playing Cinderella."

"I'm sure you weren't so sure. You were in *another* room 208, that's for sure."

HARUKI MURAKAMI
The Wind-up Bird Chronicle
(trans. Jay Rubin)

CHAPTER 2

Playing Peano

Some of the beans had taken root, and sprung up surprisingly: the stalks were of an immense thickness, and had twined together until they formed a ladder like a chain, so high that the top appeared to be lost in the clouds.

CHARLES PERRAULT
Jack and the Beanstalk, trans.
Unknown

2.1 TETRA

2.1.1 The Peano Axioms

Our first day back at school was a Friday. I wasn't in the mood to go home after classes ended, so I headed to the school courtyard. I sat on one of the benches that surrounded a small pond there. In warmer weather it was a popular place to eat lunch and hang out, but that day the January cold kept everyone else away. It felt good to me, though—the bite in the air cleared my head.

"*There* you are!" Tetra shouted from behind me. I turned to see her heading my way.

How did she find me here? I wondered. I recalled how my mother had once referred to her as my 'cute little stalker.'

She is *cute, I gotta admit.*

She was petite, curious, and full of energy. Her short bobbed hair suited her personality. She was one grade behind me, and I'd been tutoring her since we'd met. No matter where I went after classes she'd find me and ask me questions about math. Not that I really minded; I enjoyed teaching her, and we'd become good friends.

I scooted over on the bench, and she sat next to me. A sweet scent wafted my way, and I fleetingly wondered how girls managed to smell so *good.*

"I've got the latest!" Tetra said, pulling an index card from her bag.

I immediately knew it was from Mr. Muraki, one of the math teachers at our school. He was kind of a strange guy—in a good way, mind you—who had taken a liking to us. He would sometimes give us cards with problems written on them. The timing of their arrival was erratic, as was their level of difficulty. When working on them we could reference any book we pleased and ask anyone for help. Our solutions didn't have a due date, nor did they result in a grade, but we would do our best to solve them and present him with a carefully crafted report. I'm sure they were intended as something fun, a form of intellectual entertainment, but we took them very, very seriously.

"It's another weird one," Tetra said. "I can't make heads or tails of it."

The Peano axioms (as text)

PA1 1 is a natural number.

PA2 If n is a natural number, its successor n' is a natural number.

PA3 For all natural numbers n, $n' \neq 1$.

PA4 For all natural numbers m, n, if $m' = n'$ then $m = n$.

PA5 If $P(n)$ is a predicate such that

        (a) $P(1)$ is true, and
        (b) for every natural number k, if $P(k)$
            is true, then $P(k')$ is true,

then $P(n)$ is true for every natural number n.

"Huh, interesting," I said.

"There's more on the back."

I flipped the card over. Its reverse was covered in symbols.

The Peano axioms (as logic statements)

PA1 $1 \in \mathbb{N}$

PA2 $\forall n \in \mathbb{N} \left[n' \in \mathbb{N} \right]$

PA3 $\forall n \in \mathbb{N} \left[n' \neq 1 \right]$

PA4 $\forall m \in \mathbb{N} \ \forall n \in \mathbb{N} \left[m' = n' \Rightarrow m = n \right]$

PA5 $\left(P(1) \land \forall k \in \mathbb{N} \left[P(k) \Rightarrow P(k') \right] \right) \Rightarrow \forall n \in \mathbb{N} \left[P(n) \right]$

"I can't even find the problem we're supposed to solve in all that," Tetra said.

"I think he intends this one to be more like a research topic."

"So, like, we're supposed to create our own problem?"

Tetra took the card back for a closer look.

"Yeah. These are from the Peano axioms. They're pretty famous. I think he wants us to play with these axioms PA1 through PA5."

"Well, I've done my best since he gave me the card, but I don't have a single clue what all this means. Well, actually I did figure out just one thing—that 'successor n' the problem mentions."

"Nice, I can tell you read this closely. Be careful, though. It's 'successor n-*prime*,' not 'successor n.'"

"Oops, sorry! Anyway, isn't n' just $n + 1$? That's what the word 'successor' makes it sound like, at least—the next number after n."

"In effect, yeah, but—"

"Then why use n', instead of just writing $n + 1$? It feels like somebody's going out of their way to make things more confusing. And what are these Peano axioms for, anyway? This logic statement version especially is just *crazy*."

"That's only because you're seeing these for the first time, so it's a little hard to follow. I've read about these before, though, so I'm a bit more familiar with what they're saying. Instead of trying to take the whole thing in at once, how about we try reading through the axioms, one by one."

"That would be great!"

Tetra's eyes sparkled, making me smile. I loved being with people who enjoyed studying math.

"I guess we should start by talking about what Peano was trying to accomplish with all this."

"So Peano is a person?"

"A mathematician, yeah. He used these axioms to define the natural numbers."

"Wait—*define* the natural numbers?"

"Yep, that's what the Peano axioms do."

"Why on earth would you need to define them? They're just . . . *natural*. Who needs a definition? Just do like this!"

Tetra ticked off a few numbers on her fingers, accompanied by an exaggerated "ooone," "twooo," "threee."

"Think of this as Peano's way of looking deep into the natural numbers, an attempt to examine their fundamental nature. Remember, an axiom is a statement that you take to be true without a proof backing it up. A proposition is a statement that you can mathematically show to be true or false. We can use axioms PA1 through PA5 on this card to define a set $\mathbb{N}$ of all the natural numbers.

"Actually, it'll probably be better if we just forget about the natural numbers for now and just think of this set $\mathbb{N}$ as something that fulfills the requirements of the Peano axioms. That will make it easier to read through the axioms and think about what they say about $\mathbb{N}$."

Tetra started to answer, but her words morphed into a squeaky sneeze.

"But before we get started," I said, "maybe we should go somewhere warmer."

2.1.2 Infinite Wishes

Tetra and I walked down the tree-lined path that led to the student lounge. I noticed she was trotting to keep up with me, so I slowed my pace to accommodate.

"You know how there's all these fairy tales about getting three wishes?" she asked.

"Like from genies that appear when you rub a lamp?"

"Those exactly. If I ever meet a genie, I know what my third wish is going to be for—three more wishes!"

"I guess that would also be your sixth wish, and your ninth wish, and your twelfth wish, too."

"Yup, yup, and yup!"

"Well before you get to your third wish, you'd better make sure that meta-wishes are allowed."

Tetra came to a stop. "What's a meta-wish?"

"A wish about wishes."

"Oh, I've studied that word 'meta'! It's from the Greek word for 'beyond.'"

"Hey, wait," I said, stopping. "There's a better way—just make your first wish be for infinitely many wishes. Then you only need to make the one meta-wish."

Tetra cocked her head and considered that.

"But then you're making infinitely many demands in just one breath. That seems kind of selfish, somehow."

"I guess."

We resumed walking along the path.

"So what would your first non-meta-wish be for?" I asked.

"You—ah!" Tetra froze, then coughed. *"You* know, the usual," she said.

She looked straight down the path and continued toward the lounge at a much brisker pace.

2.1.3 Peano's Axiom PA1

The student lounge was a popular hangout, but it was quiet that day. We bought coffee from a vending machine and sat at a table, next to each other so we could share a notebook.

I opened to a fresh page and began by writing the first axiom.

Peano's axiom PA1

1 is a natural number.

$$1 \in \mathbb{N}$$

"Do you understand the $1 \in \mathbb{N}$ here?" I asked.

"That's the symbol for 'element of a set,' isn't it? So this says that 1 is an element of a set named $\mathbb{N}$."

"That's exactly right."

$$1 \in \mathbb{N} \qquad \text{1 is an element of set } \mathbb{N}$$

"You'll also hear people say '1 is in $\mathbb{N}$,' or '1 is a member of $\mathbb{N}$,' or '1 belongs to $\mathbb{N}$.' They all mean the same thing."

"Got it!"

"We can picture $1 \in \mathbb{N}$ like this."

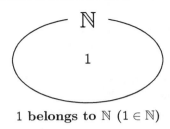

1 belongs to N (1 ∈ N)

"Right!"

"Okay, then. On to axiom PA2."

2.1.4 Peano's Axiom PA2

"This next axiom introduces something called a 'successor,'" I said.

Peano's axiom PA2

If n is a natural number, its successor n' is a natural number.

$$\forall n \in \mathbb{N} \left[n' \in \mathbb{N} \right]$$

"And it just means $n + 1$, right?"

"In the end it does, I guess, but PA2 itself doesn't go quite that far."

"I'm not sure what 'doesn't go that far' means."

One of Tetra's best habits was never going ahead when she wasn't sure she understood everything. She often complained that it took her twice as long to learn things, but I think she was comparing herself to people who were going too fast.

"Let's take a careful look at PA2," I said. "It says, 'If n is a natural number, its successor n' is a natural number.' We have to use n' instead of $n + 1$, because we're still in the process of defining what the natural numbers are. If that's really where we're heading, that is—we could instead be defining the *odd* natural numbers, in which case $n + 2$ would be the successor. Either way, without numbers, it's way too early to define something like addition."

"Oh."

"Eventually, yeah, we'll find that n' is equivalent to $n+1$. But not quite yet. For now, let's avoid the preconceived notion that they're the same thing."

"Okay, that makes sense—we haven't defined what $n + 1$ means, so we can't use it yet, right?"

"Exactly! So do you see what it is that axiom PA2 wants to say? Specifically, what it's saying about elements of the set of natural numbers $\mathbb{N}$?"

"Hmm, lemme see... It says that for every natural number n, n' is a natural number too. So I guess it's saying that since 1 is a natural number, so is 2."

"Hang on. Where'd that 2 come from?"

"What do you mean? We know that 1 is a natural number, so when you add 1 to that you get 2—"

"But you're *adding*."

"Oops! We can't do that yet, can we..."

"You're right that 1 is a natural number—axiom PA1 assures us of that. What PA2 is saying is that $1'$ is a natural number, too. It's a subtle distinction, but we can't talk about 2 yet, just $1'$."

"Wow, I didn't realize we have to take things quite so literally."

"Such is the price of precision. We can say that $1' \in \mathbb{N}$, though, like this. First, PA1 tells us that 1 is a natural number."

$$1 \in \mathbb{N}$$

"PA2 says that for any natural number n, n' is a natural number too, so we get this."

$$n' \in \mathbb{N}$$

"Now we can replace n with 1, which tells us that $1'$ is a natural number."

$$1' \in \mathbb{N}$$

"I know it seems picky, but it's important to stick to what the axioms allow for."

"I think I'm starting to get the hang of it, though," Tetra said. "You have to pretend like you know absolutely nothing. Even when you see where things are headed, you pretend like you don't, and just follow the rules. That which isn't specifically allowed is forbidden."

"A perfect description! 'That which is allowed' is what the axioms say you can do, and you can't do anything else. If it isn't defined, it doesn't exist."

"And in this case, using PA1 and PA2 tells us that now 1 and 1′ are in the set of natural numbers. Let me draw it this time!"

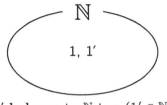

1′ **belongs to** $\mathbb{N}$ **too** $(1' \in \mathbb{N})$

"Well done."

"So what's the upside-down 'A' in the logic statement version of this?"

Tetra pointed at PA2 on the back of the card.

"You read that as 'for all' or 'for every.' The basic pattern of those statements is something like this."

$$\forall n \in \mathbb{N} \left[\text{some proposition in } n \right]$$

"You can think of this as meaning 'for all elements n in $\mathbb{N}$, some proposition in n is true.' You can also write this using the 'implies' symbol, an arrow."

$$\forall n \left[n \in \mathbb{N} \Rightarrow \text{some proposition in } n \right]$$

"If it's clear from context that n is an element of $\mathbb{N}$, you can even abbreviate."

$$\forall n \left[\text{some proposition in } n \right]$$

"When you read a lot of math books you might see other variations, but they all mean basically the same thing."

"Got it!" Tetra said, pumping a clenched fist. "Now, on to axiom PA3!"

"Not so fast. We still haven't finished with PA2."

2.1.5   Getting Bigger

"We've learned that $1'$ belongs to the set $\mathbb{N}$," I said. "In other words, $1'$ is a natural number. So what happens if you apply axiom PA2 to $1'$?"

Axiom PA2:
  If n is a natural number, its successor n' is a natural number.

Tetra placed a finger on her chin. "Do you get ... the successor of a successor?"

"You do indeed. We know that $1'$ is a natural number, which means its successor $1''$ must be too."

$$1'' \in \mathbb{N}$$

"And you can keep doing that? Just add more prime symbols to make more successors?"

"Sure."

$$1 \in \mathbb{N}, \ 1' \in \mathbb{N}, \ 1'' \in \mathbb{N}, \ 1''' \in \mathbb{N}, \ 1'''' \in \mathbb{N}, \ \ldots$$

"Makes sense, I guess. If the natural numbers are going to be $1, 2, 3$ and on and on, we'll need to be able to make lots of these successors, won't we."

"And axioms PA1 and PA2 let us do that. They've made our set $\mathbb{N}$ much bigger now."

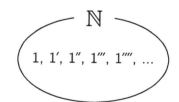

$1, 1', 1'', 1''', 1'''', \ldots$ **belong to** $\mathbb{N}$

"Neat!"
"Now we can write set $\mathbb{N}$ like this."

$$\mathbb{N} = \left\{ 1, 1', 1'', 1''', 1'''', 1''''', \ldots \right\}$$

"And that defines the natural numbers?"

"Not yet. We've still got some more axioms to get through first."

"Oh, right. Now that I'm getting into the swing of things, this is fun!"

· You can use anything that is given as an axiom.

· You can use anything you've logically derived from the axioms.

· You can use axioms repeatedly.

· By doing this, we defined the set $\mathbb{N}$.

"Good summary. Of course, what we're really hoping is that $\mathbb{N}$ will end up being the set of all natural numbers. But so far all we have is $\mathbb{N}$ as a set of 1 and lots of successors."

"Which we're pretending we don't know are the natural numbers!"

"Well put."

"It's pretty amazing that you can use just two simple rules to create an infinity of numbers! It's kinda like placing two mirrors facing each other!"

"Not so fast..."

"Uh-oh." Tetra frowned. "I think I know what's coming..."

2.1.6 Peano's Axiom PA3

"You're going to say we can't claim to have created infinitely many numbers, right?" Tetra said.

I nodded. "Yep."

"But why not? I mean, we can add as many prime symbols as we like, right? There's nothing limiting how many we can add, is there?"

"No, but that still doesn't let us claim to have created infinitely many elements."

"But...but—we created this set $\{1, 1', 1'', 1''', 1'''', \ldots\}$, right?"

"We did."

"Don't those dots mean 'and so on and so on, forever and ever'?"

"They do. But they don't guarantee that 1 and 1' and 1'' and the others are all *different* numbers."

"Hang on. 1 and 1' are different numbers, right?"

"We don't know. Nothing in axioms PA1 or PA2 allow us to say that $1' \neq 1$."

"Wow, now *that's* some serious casting aside of preconceived notions."

"Not for long, though. Axiom PA3 is going to clear things up for us."

I pointed at PA3 in the notebook.

"Good old Peano! I knew he'd come through!"

Peano's axiom PA3

For all natural numbers n, $n' \neq 1$.

$$\forall n \in \mathbb{N} \left[n' \neq 1 \right]$$

"So this lets us say that $1' \neq 1$?" Tetra asked.

"Sure. Just let the n in $n' \neq 1$ be 1."

"Which we know we're allowed to do, since we know 1 is a natural number." Tetra sighed. "I need to get more comfortable with this 'for all natural numbers' thing," she said. "It goes against my instincts somehow, to just think about whether n is a natural number or not, ignoring everything else about it. I think this 'no questions asked' attitude is what gives me such a rough time with conditions and logic and all."

"Interesting. You're just the opposite of Yuri—she seems to like how the 'no questions asked' part of logic lets her jump to an absolutely right answer. I totally get where you're coming from, though. You and I are a lot a like. We're worriers."

Tetra looked away, but not in time to hide her blush.

"Sorry I'm always saying all this weird stuff," she said.

"Not at all. I learn a lot from it."

Tetra faced me again. I was relieved to see her usual smile had returned.

2.1.7 How Small is Small?

As I sipped my now-cold coffee, Tetra raised her hand.

"I have a question about axiom PA3," she said.

"What's that?"

"Does it say that 1 is the smallest number?"

Axiom PA3:
   For all natural numbers n, $n' \neq 1$.

"Mmm...yes and no."

"Huh?"

"It says that 1 plays a very special role, but we can't call that being the 'smallest' number yet. Do you see why?"

Tetra scrunched her face in thought, and I sat back to let her mull on this. I realized how unnaturally quiet the lounge was that day. It felt odd without the usual hum of students talking and the muffled screeches and squawks from members of the band practicing upstairs.

Tetra was biting her lower lip, which—not for the first time— made me imagine her as a giant squirrel. I bit my own lip to keep from laughing.

"I'm not seeing it," she said, shaking her head.

"Don't forget the big picture here—we're trying to define the set of the natural numbers. We aren't done with that yet, so we can't bring in any of the tools we take for granted when we normally do math."

"I'm not even sure what that *means*."

Tetra was clearly becoming distressed.

"It means we don't even have a concept of 'small' yet. We have no definitions of 'smaller' or 'larger,' so we have nothing to apply to determine if 1 is the 'smallest' number."

"Oh. Oh, okay. 'If it isn't defined, it doesn't exist,' right?"

"Right. Now let's get back to our list. Just two more axioms to go."

2.1.8   Peano's Axiom PA4

"Let's see," I said, "next was..."

"Right here," Tetra said, pointing to the notebook.

Peano's axiom PA4

For all natural numbers m, n, if $m' = n'$ then $m = n$.

$$\forall m \in \mathbb{N} \; \forall n \in \mathbb{N} \left[m' = n' \Rightarrow m = n \right]$$

"Take a shot at unpacking that," I said.

"Well, let's see ... I think I understand what these mean, both the text version and the logic statement. You said this arrow means 'implies,' right?"

"Right."

"Then I understand what it means to say that $m' = n'$ implies $m = n$. You're saying that if m' and n' are the same number, then m and n are the same number too. I guess what I don't understand is what this has to do with defining the natural numbers."

"The goal here isn't quite as obvious as with the first three axioms, is it."

"But what it's saying seems kinda obvious, doesn't it? I mean, $m' = n'$ is gonna be basically $m + 1 = n + 1$, so how could m and n possibly *not* be the same number?"

"There you go again, abusing the axioms. You're looking at things backwards, and letting the meaning of the word 'successor' mislead you. You've noticed that *eventually* we're going to find that m' means $m + 1$, and that's why this axiom feels so obvious to you. Those preconceived notions are creeping back in."

"Hard habit to shake, I guess."

"Yeah. What this axiom is really doing is setting up successors so that if $m' = n'$, then $m = n$. In other words it's *defining* this 'prime' operator for finding successors, giving it the property that if $m' = n'$, then $m = n$."

"Oh, I see. So the prime is an operator. Ugh, there's something about this that makes my head hurt. It's like we're doing math, but we aren't. It feels like I have to use a different part of my brain."

Tetra held her head in her hands.

"Maybe, but doing something like this is necessary to avoid a loop."

"What kind of loop?"

"Well, the way I picture it is like this. We've already found that $\mathbb{N}$ is the set $\{1, 1', 1'', 1''', 1'''', \ldots\}$, right? Think of using the prime operator as a way of walking along those elements. I see it as a chain like this."

$$1 \longrightarrow 1' \longrightarrow 1'' \longrightarrow 1''' \longrightarrow 1'''' \longrightarrow \cdots$$

"So you're starting with 1 and going in order, from successor to successor."

"Right. But imagine what would happen if along the way we found that, say, $1'$ and $1'''''$ were equivalent."

"Uh... what would happen?"

"We wouldn't be on a straight path any more. We'd be walking in circles!"

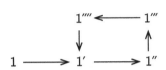

$1'$ and $1'''''$ being equivalent creates a loop

"Oh, okay... Because when you get to $1'''''$, that sends you back to $1'$."

"Which isn't what we're after. We want to keep moving forward, without getting stuck in loops along the way."

"Whoa!"

Tetra's hand shot out, grasping my arm. Her expression was deadly serious.

"Wait, wait, wait a minute, I'm starting to get this. I think I understand why you said I was thinking backwards. The properties of n' are all represented by the axioms. Right, right..."

Tetra continued, apparently oblivious of the fact that she was shaking my arm as she spoke.

"That's it," she continued. "We want to say that 1 is a natural number, so we use PA1, $1 \in \mathbb{N}$. We want to say that every natural number has a successor, so we use PA2, $n' \in \mathbb{N}$. And since there's no natural number smaller than 1—no, wait. We can't say 'smaller' yet... Um, since we want to say that there's no natural number that has 1 as its successor, we use PA3, $n' \neq 1$. And finally, we want to say that we can keep on going from successor to successor, onward and onward with no turning back. *That's* why we use this fourth axiom!"

"Tetra, somewhere out there Peano is smiling right now—you're seeing the image of the natural numbers that he wanted to describe."

"A message straight from Peano. Cool."

Tetra suddenly let go of my arm and stood from her seat, her face flushed and eyes wide. I realized that she wasn't looking at me, but over my shoulder. I turned to see what had surprised her.

There stood Miruka, arms crossed and a smile on her face.

2.2　Miruka

Miruka was one of my classmates. She was pretty good at math, the way dolphins are pretty good at swimming. She had metal-framed glasses, long black hair, and a tongue that could slice you to ribbons. I was sure there were things going on in her head that weren't related to math, but what, I couldn't say. Since we first met, she'd been something of an enigma.

"So this is where you two have been hiding," Miruka said, striding over to our table. "Did you get a card, Tetra?"

"I...I did," Tetra stammered, fishing for the card. She held it out to Miruka.

Miruka took the card, and Tetra deflated into her chair.

"Peano arithmetic, huh." Miruka said, comparing the two sides of the card.

"Is that what all this is?" Tetra asked.

Miruka pushed her glasses up her nose.

"PA1 through PA5 here are the Peano axioms. Once you've used those to develop the set $\mathbb{N}$, you can go on to define predicates, then addition and multiplication. That's when you get to the arithmetic part. Not there yet, I suppose?"

"Not quite," I said. I gave Miruka a rundown of what we'd covered.

Miruka moved behind me. She leaned over my shoulder to take a closer look at our notes, enveloping me in an invisible, citrus-scented cloud. Her hair brushed my cheek. She placed her hand on my shoulder.

So warm...

"Hmph. Loops, huh? I guess that's not wrong."

She stood, closed her eyes, and the air regained its previous chill. Tetra and I remained silent, waiting.

"Not wrong, but not good enough," Miruka said, opening her eyes.

"No?" I said, feeling a little panicked. "But you see how if we didn't have PA4—or rather, if we didn't have *some* axiom saying $m' = n' \Rightarrow m = n$—then the path you'd follow if you traced along successors could end up looping, right?"

"That's all fine. But it isn't really looping that PA4 prevents, it's more like merging. Of course no merging means no looping, too."

"What do you mean, merging?" I asked.

"Easier to explain with a graph."

Miruka flicked her hand at Tetra in a get-out-of-that-seat-or-be-destroyed gesture. Tetra froze for a moment, but soon stood and woodenly made her way to the other side of the table. Miruka sat down—

Next to Tetra?

"Look at it this way," Miruka said. "With just the first three axioms, you could end up with a structure like this."

Miruka snagged the pencil from my hand and drew a new graph.

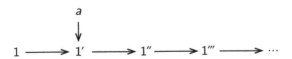

PA1, PA2, and PA3 alone allow merging

"Doesn't that look more like merging?" Miruka asked, raising an eyebrow.

"Hang on now," I said. "Where did that a come from? You certainly didn't get there starting from 1."

"Take another look at the axioms. Where in PA1, PA2, or PA3 does it say you have to be able to reach all elements by starting at 1? Until you show me otherwise—using just these three axioms—I claim that I can introduce an element a that isn't reachable from 1. You're right that PA4 prevents loops, but it also prevents elements like a from jumping in."

"Hey Miruka...," Tetra said. "If PA4 forbids merging like that, then wouldn't that mean we don't need PA3, that $n' \neq 1$?"

"No," Miruka said. "If you just used PA1, PA2, and PA4, the natural numbers could look something like this."

Miruka drew a new graph.

PA1, PA2, and PA4 alone allow a structure like this

"Granted, there's no merging," Miruka said. "But that's not the structure you want the natural numbers to have, is it?"

"You're right," I said. "We could end up with the natural numbers not just heading off into some infinite future, but also appearing from out of some infinite past."

"Mr. Peano really thought this stuff through, didn't he," Tetra said.

"He's not done yet," Miruka said. "There's still one more axiom to cover."

2.2.1 Peano's Axiom PA5

Peano's axiom PA5

If $P(n)$ is a predicate such that

    (a) $P(1)$ is true, and

    (b) for every natural number k, if $P(k)$ is true, then $P(k')$ is true,

then $P(n)$ is true for every natural number n.

$$\left(\underbrace{P(1)}_{(a)} \overset{\text{and}}{\wedge} \underbrace{\forall k \in \mathbb{N} \left[P(k) \overset{\text{implies}}{\Rightarrow} P(k') \right]}_{(b)} \right) \Rightarrow$$

$$\forall n \in \mathbb{N} \left[P(n) \right]$$

"Axiom PA5 introduces something new," Miruka said, "a 'predicate' in a natural number n."

"Is that anything like a predicate in grammar?" Tetra asked.

"Not really. A predicate in n becomes a mathematical proposition when you assign some specific value to n. The predicate is called $P(n)$ in this axiom, but that's just an arbitrary name.

"PA5 describes a method of proving that $P(n)$ holds for all natural numbers n—mathematical induction. It's quite significant, actually, that induction appears along the way toward a definition of the natural numbers. It's an indication of how closely their natures are related."

"How, specifically?" I asked.

"Well, assume there were a finite number of natural numbers. Just $1, 2, 3$, for example. Then you'd just have to show that $P(1)$, $P(2)$, and $P(3)$ were valid to prove some proposition $P(n)$ for all natural numbers n."

"Oh!" Tetra said. "But there are infinitely many natural numbers, so you can never prove anything taking them one at a time!"

"Which is exactly why we need a tool like mathematical induction—so that we can make claims about the infinite in a finite space. Peano's axiom PA5 gives us that."

"Um, Miruka?" Tetra said, looking down. "I've heard of mathematical induction, studied it in class even, but, uh..."

"A quick primer, then," Miruka said with a wink.

2.2.2 Mathematical Induction

"There are two steps in mathematical induction," Miruka began. "You set up the foundation by proving that some proposition P(1) holds. After that, you show that if P(k) holds, then P(k + 1) holds too. If you can do both, then you've proved that P(n) holds for all natural numbers n. Make sense?"

"I think so..." Tetra said, nodding slowly.

"Let's run through an easy one just to be sure. One caveat, though—we have to assume that the Peano axioms have defined the natural numbers, and that we've also defined the basic arithmetic functions."

Problem 2-1 (Odd sums and squares)

Show that the following holds for any natural number n:

$$1 + 3 + 5 + \cdots + (2n - 1) = n^2$$

"Okay, I'll give it a shot," Tetra said. "So the first step in mathematical induction is—"

"Wait!" Miruka said, slapping the table. "Start with an example. *Always* start with an example. It's foolish to do anything else."

"Oh, yeah! 'Examples are the key to understanding,' right?" Tetra shot me a knowing glance.

"Then how about this?"

$$1 = 1 = 1^2 \qquad \text{for } n = 1$$
$$1 + 3 = 4 = 2^2 \qquad \text{for } n = 2$$
$$1 + 3 + 5 = 9 = 3^2 \qquad \text{for } n = 3$$

"Okay, looks like it works," Tetra said. "For the first few values of n, at least."

She nodded again, this time with more conviction.

"I noticed something writing this out," she continued. "This expression, $1 + 3 + 5 + \cdots + (2n - 1)$, is always going to be a bunch of odd numbers added together, right?"

"An excellent thing to notice," Miruka said, holding up an index finger. "Examples hold compressed truths. It's human nature to unconsciously search for patterns in examples, and thereby find concise descriptions. Like saying $1 + 3 + 5 + \cdots + (2n - 1)$ instead of 'a bunch of odd numbers added together.'

"Anyway, back to proving this statement. There's several ways to do it. Try to think of one that uses mathematical induction, Tetra."

"Wow, okay. Let's see ... "

Tetra drew a deep breath and clutched her pencil before continuing.

"We can define a predicate $P(n)$ on a natural number n like this."

Predicate $P(n)$:
$$1 + 3 + 5 + \cdots + (2n - 1) = n^2$$

"Then I guess this is all we need for the first part of the proof."

Step (a):
We start by showing that $P(1)$ holds:

Proposition $P(1)$:
$$1 = 1^2$$

This is a true statement, so $P(1)$ holds.

"The second step is a little trickier, but I think I see how to get started, at least."

Step (b):
   Next we want to show for all natural numbers k that if $P(k)$ holds, then $P(k+1)$ holds too. Suppose that $P(k)$ holds. That means the following is true:

(Assumed) $P(k)$:
$$1 + 3 + 5 + \cdots + (2k - 1) = k^2$$

"So now I have to use this to show that $P(k+1)$ must be true too. Let's see ... $P(k+1)$ looks like this."

We want $P(k+1)$:
$$1 + 3 + 5 + \cdots + (2\underline{(k+1)} - 1) = \underline{(k+1)}^2$$

"I guess the only thing that changes is the k's become $(k+1)$'s ... Anyway, what we want to do is derive $P(k+1)$ from $P(k)$. In other words, we want to start with

$$1 + 3 + 5 + \cdots + (2k - 1) = k^2,$$

and rearrange things so that the left side of this becomes the left side of $P(k+1)$. Maybe if we add $(2(k+1) - 1)$ to both sides ..."

Add $(2(k+1)-1)$ to both sides of $P(k)$, giving the following:

$$1 + 3 + 5 + \cdots + (2k-1) + \underline{(2(k+1)-1)} = k^2 + \underline{(2(k+1)-1)}$$

remove parentheses

$$= k^2 + \underline{2(k+1)-1}$$

remove more parentheses

$$= k^2 + \underline{2k+2}-1$$

combine constants

$$= k^2 + 2k + \underline{1}$$

rewrite as a square

$$= \underline{(k+1)^2}$$

"Okay, so we've ended up with this."

$$1 + 3 + 5 + \cdots + (2k-1) + (2(k+1)-1) = (k+1)^2$$

"And that's what $P(k+1)$ looks like!"

The result is a form equivalent to $P(k+1)$. In other words, we have derived $P(k+1)$ from $P(k)$, thus completing the proof of this step.

We have shown that steps (a) and (b) both hold, and thus have proven by mathematical induction that $P(n)$, in other words

$$1 + 3 + 5 + \cdots + (2n-1) = n^2,$$

holds for all natural numbers n. □

"Q.E.D.!" Tetra half shouted.

"Very good," Miruka said.

I sat with my jaw hanging, amazed at how well she had done.

"Tetra, I thought you said you didn't know how to do these?"

"Well, I can go through the motions, treating induction like a template I'm supposed to follow. I did study it in class, after all. But I don't feel like I really, *really* understand it."

"What in particular bothers you?" Miruka asked.

"Well, like in this proof I just did—I said 'Suppose that P(k) holds,' right? But how can I suppose that? I mean, the whole point of all this is to prove that P(n) is true, right? So it seems kinda like cheating to just go ahead and assume that it is. Yeah, I can write a proof by induction okay. What's harder is convincing myself I've actually proved anything."

Miruka raised an eyebrow at me, handing off the baton.

"That's an excellent point, Tetra," I said. "Maybe it would help if you thought of induction as something like knocking over dominoes."

Tetra blinked. Twice.

"Like when you line them up and, *brrrrt*, make them all topple over in a chain reaction?"

"Right. What you're setting out to prove is that eventually all the dominoes will fall. Proving the first step in an induction proof is like showing that the first domino falls. In the second step, you show that if domino number k falls, then domino (k + 1) will, too. So what you're saying is, 'if one domino falls, then so will the one coming after it.' You see the difference in these, right?"

- Saying that if a domino falls, the next one will, too.

- Saying that a domino falls.

"Well sure," Tetra said. "One's an 'if,' the other's an 'is.'"

"Yep," I nodded. "The same thing's going on in a proof by induction, though the difference might not be as obvious. Can you see which of these is the second induction step?"

(1) For all natural numbers k, if P(k) holds then P(k + 1) holds too.

(2) If P(k) holds for all natural numbers k, then P(k + 1) holds too.

"Oh, I see!" Tetra said. "Mathematical induction uses the first one, right? But I'd thought it assumed the second one. Things make so much more sense this way!"

Answer 2-1 (Odd sums and squares)

We let predicate $P(n)$ be defined as

$$1 + 3 + 5 + \cdots + (2n - 1) = n^2,$$

and use mathematical induction to demonstrate that it holds for every natural number n.

(a) $1 = 1^2$, so $P(1)$ holds.

(b) Supposing that $P(k)$ holds for some natural number k, we obtain the following:

$$1 + 3 + 5 + \cdots + (2k - 1) = k^2$$

After adding $(2(k+1) - 1)$ to both sides and cleaning up, we obtain

$$1 + 3 + 5 + \cdots + (2k - 1) + (2(k+1) - 1) = (k+1)^2.$$

In other words, $P(k + 1)$ holds.

Items (a) and (b) both hold, so by mathematical induction we have shown that $P(n)$ is true for every natural number n. $\square$

2.3   Infinite Walks

2.3.1   The Finite and the Infinite

I looked outside and was surprised at how dark it had become. We left the student lounge and headed toward the train station, walking single file along the narrower roads—Tetra, me, Miruka. No one talked for a while, so my mind was left to wander.

I thought about how we make our way, one step at a time. About how hard it is to see the path before us, and to know where we're headed. How we live our lives one day at a time, with no way of knowing what our futures hold. 'That's the way the world works,' Miruka had once said to me. 'You never know what's going to come next.'

We were wandering through a twisting maze, leaving memories as footsteps.

Walking with Tetra through spring rain, limping through beams of amber light alongside Miruka. So many walks, past, present, and future.

Tetra looked back and said, "So it just takes five axioms to define the natural numbers, huh? Pretty awesome."

"It is," I said, nodding.

"Then again, there are axioms and there are axioms. That PA5 packs quite a punch."

"That's what it takes to capture the infinite within the finite," Miruka said. "I think it's wonderful."

She cocked her head, considering this.

"Then again, there are infinities and there are infinities," she continued, mirroring Tetra. "The ones we've played with can be cast into specific forms, bound up by certain limits, described by mathematical notations. There are other infinities that lie out of our reach, refusing to fall into patterns we can perceive."

2.3.2 Dynamic vs. Static

"I guess we can say that Peano perceived the natural numbers, though," I said. "Well enough to capture them with the 'next step' form of successors, at least."

"Who'd have thought that 'one more step' could be so powerful," Tetra said.

"Especially when applied to the step-by-step approach of proof by mathematical induction," I agreed.

"You think induction proofs happen step-by-step?" Miruka asked. "I'm not sure things are quite so dynamic."

"How's that?" I asked.

"Well, it's fine to think of things in terms of dominoes, events happening one after another. It's a nice metaphor for learning. But it's too shallow for deeper understanding. Don't forget—induction gives proofs regarding *all* the natural numbers. You aren't making assertions about them one at a time, you're dealing with the whole lot as a set. That feels more static to me."

"Huh. Now that you put it that way..."

"This feels familiar," Tetra said. "It's like when you were teaching me about sequences and used expressions like $a_n < a_{n+1}$. If you're going through the sequence and comparing terms in pairs, it feels like the terms are getting bigger. But if you say that $a_n < a_{n+1}$ for all natural numbers n, that does feel more static, doesn't it."

"Whichever way you look at it," Miruka said, "there's a bigger picture to be aware of. You can use Peano's axioms to define the set of natural numbers. The whole thing is based on just sets and logic. Peano's end goal was to use just sets and logic to develop a foundation for all of mathematics."

"*All* of mathematics, from just sets and logic?" I said.

"Light's changing!" Tetra shouted, dashing for the crosswalk. The crossing signal turned red just as Miruka and I reached it, leaving us on the other side of the street. Tetra laughed and waved at us as we waited for the light to change again. I smiled and waved back.

"Back in the lounge," I said, turning to Miruka, "I was kinda surprised when you sat next to Tetra."

Miruka didn't respond. I gave in to the awkward silence and turned back to face the street.

When she finally spoke, her eyes never left the crossing signal.

"I can see your face better when I sit across from you."

"You—oh."

"Light's green," she said, stepping into the crosswalk.

2.4   YURI

2.4.1   What's Addition?

"Peano arithmetic, huh." Yuri said. We were hanging out in my room, as we did most Saturdays. I'd given her a rundown of our math talk from the previous day. "I like it."

"What do you like about it?" I asked.

"The *bam*, *bam*, *boom* of the whole thing. *Bam*, here's yer 1. *Bam*, here's yer successor function. *Boom*, behold the natural numbers. What, you're worried about loops and mergers? Nah, I've gotcha covered. Airtight and outta sight. Good job, Peano."

"I'm sure he'd be glad you approve."

"But you guys never got around to defining addition?"

"Ran out of time. It's not that hard, though."

Addition axioms

ADD1 For any natural number n, $n + 1 = n'$.

ADD2 For any natural numbers m, n, $m + n' = (m + n)'$.

Yuri squinted at what I'd written.

"You sure you got this right?" she asked.

"Yep. This defines the $+$ operator."

"Show me how $1 + 2 = 3$, then."

"Okay, but we have to use $1 + 1'$ in place of $1 + 2$."

"Oh, right. We don't have a 2 yet ..."

$$
\begin{aligned}
1 + 1' &= (1 + 1)' &&\text{axiom ADD2, with } m = 1, n = 1 \\
&= (1')' &&\text{axiom ADD1, with } n = 1 \\
&= 1'' &&\text{remove the parentheses}
\end{aligned}
$$

"See?" I said. "That gives us $1 + 1' = 1''$. Now just give $1'$ and $1''$ the names '2' and '3,' and you get $1 + 2 = 3$. Convinced?"

"Hmm, maybe that one was too easy. Try $2 + 3 = 5$."

"No problem."

$$
\begin{aligned}
1' + 1'' &= (1' + 1')' &&\text{axiom ADD2, with } m = 1', n = 1' \\
&= ((1' + 1)')' &&\text{axiom ADD2, with } m = 1', n = 1 \\
&= (((1')')')' &&\text{axiom ADD1, with } n = 1' \\
&= 1'''' &&\text{remove the parentheses}
\end{aligned}
$$

"It works pretty much the same as before, but this time we get $1'+1'' = 1''''$. Give $1', 1'', 1''''$ the names $2, 3, 5$, and you have $2+3 = 5$."

"Okay, I'm convinced," Yuri said, nodding. "I'm going to have to get used to the idea of naming numbers, though."

2.4.2   More Axioms Needed?

"So what's the deal with Tetra?" Yuri asked, looking at the induction proof Tetra had written in my notebook. "How can she whip this stuff out and still think she sucks at math?"

"Yeah, funny isn't it. I think she just hasn't realized how much progress she's made. She's tenacious when it comes to studying, and it's paying off. I wish *somebody* I could name did the same."

Yuri scowled and stuck her tongue out at me.

"I'm aiming for the double punch of brains and beauty, like Miruka," she said, her expression softening when she mentioned her idol. "I wonder what her secret is . . . "

"I'm pretty sure she studies like a fiend, too."

"Somehow that's hard to imagine."

Yuri turned back to the notebook, closely examining everything we'd written.

"Mathematical induction, huh? This is kinda neat. So the sum of n odd numbers equals n squared? Cool . . . "

After a moment's reflection, Yuri narrowed her eyes and looked up at me.

"Something's strange here," she said.

"What's that?"

"Where'd this 'equals' thing come from?"

"Huh?"

"Well you made such a big fuss about not being able to use the $+$ operator until it was defined. How come the $=$ operator gets a free pass?"

I looked up from the book I'd been browsing through and considered this.

"You're right," I concluded, wincing.

Yuri grinned and continued.

"While we're at it, you haven't defined the $\overset{\text{in}}{\in}$ operator, either!"

"Uh, yeah. That, uh ... "

"Practically none of these symbols have definitions! Upside down A's? Arrows? All undefined!"

Yuri leaned toward me with a leer.

"Time to cough up some more axioms, dude."

In the last analysis, *every* property of the integers must be proved using induction somewhere along the line, because if we get down to basic concepts, the integers are essentially *defined* by induction.

DONALD KNUTH
The Art of Computer Programming,
Vol. 1

Yuri —

Here's some more axioms for you.

Satisfied now? =P

THE PEANO AXIOMS

$$1 \in \mathbb{N}$$

$$\forall n \in \mathbb{N} \left[n' \in \mathbb{N} \right]$$

$$\forall n \in \mathbb{N} \left[n' \neq 1 \right]$$

$$\forall m \in \mathbb{N} \ \forall n \in \mathbb{N} \left[m' = n' \Rightarrow m = n \right]$$

$$\left(P(1) \overset{and}{\wedge} \forall k \in \mathbb{N} \left[P(k) \Rightarrow P(k') \right] \right) \Rightarrow \forall n \in \mathbb{N} \left[P(n) \right]$$

ADDITION AXIOMS

$$\forall n \in \mathbb{N} \left[n + 1 = n' \right]$$

$$\forall m \in \mathbb{N} \ \forall n \in \mathbb{N} \left[m + n' = (m + n)' \right]$$

MULTIPLICATION AXIOMS

$$\forall n \in \mathbb{N} \left[n \times 1 = n \right]$$

$$\forall m \in \mathbb{N} \ \forall n \in \mathbb{N} \left[m \times n' = (m \times n) + m \right]$$

INEQUALITY AXIOMS

$$\forall n \in \mathbb{N} \left[\neg (n < 1) \right] \quad \text{(n is never less than 1)}$$

$$\forall m \in \mathbb{N} \ \forall n \in \mathbb{N} \left[(m < n') \iff (m < n \overset{or}{\vee} m = n) \right]$$

CHAPTER 3

Galileo's Doubts

> On the contrary, it is words that are vague. The reason why the thing can't be expressed is that it's too *definite* for language.
>
> C.S. Lewis
> *Perelandra*

3.1 SETS

3.1.1 A Rude Awakening

"—up! Wake up! Wake *uuup!*"

I jolted upright and looked about in a blurry panic, sure I'd find myself in the middle of a fire or a zombie apocalypse. But it was just Yuri.

"Stop sleeping at your desk," she said.

"I was just...thinking really hard," I said, rubbing my eyes.

"Do you always drool when you think so hard?"

My hand shot across my mouth.

"Ha! Gotcha!" Yuri laughed.

It was the weekend, and as usual Yuri was going to hang out while I tried to study. Or so I'd assumed—that day *she* was the one who wanted to do math.

"Okay, today you're going to teach me about sets."

"Why sets?"

"My math teacher brought them up at the end of class last week, said they were fun to play with. You were talking about sets the other day, too, right? So I figure they must be important."

"You could say that."

"My teacher said something about sets being a collection of things, but not like a collection of comic books or Pez dispensers. Then everybody got sidetracked talking about the stuff they collect and everything fell into chaos until the bell rang."

"Ah, those were the days," I said.

"So anyway, tell me what sets are all about. If you can do it without falling asleep again."

"Nothing like math to keep me awake," I said, reaching for a notebook.

3.1.2 Extensional Definitions

"Give me some multiples of 2," I said. "We can just stick to the natural numbers for now."

"You mean like $2, 4, 6$?" Yuri replied.

"Right. Just start writing multiples of 2."

"Easy enough."

$$2, 4, 6, 8, 10, 12, 14, 16$$

"That's enough," I said, reaching over to add ellipses to what she'd written. "We're going to call this the set of all natural number multiples of 2. To show that it's a set, we'll add some curly braces."

$$\{2, 4, 6, 8, 10, 12, 14, 16, \dots\}$$

"But all we've done is list some numbers."

"Sure, following the rules for indicating a list. Here's all of them."

· Separate individual elements with commas.

· It doesn't matter what order you put elements in.

· If there are infinitely many elements, indicate that with ellipses.

· Surround the whole thing with curly braces.

"Ugh. I hate writing curly braces. Mine always come out funny."

"You'll get used to it."

"So a set is just a bunch of numbers?"

"A bunch of *somethings*. They don't have to be numbers."

"And I just have to write a bunch of somethings and put braces around them? This is easy."

"Try to call them 'elements,' though, or 'members.' Not somethings."

"What's an element?"

"Just a thing that belongs to the set. Like, 10 is an element of this set we just made. You show that using this symbol."

$$10 \in \{2, 4, 6, 8, 10, 12, 14, 16, \ldots\}$$

"You math people sure do love your symbols," Yuri said, shaking her head.

"Symbols rock," I said. "They save so much writing."

Yuri shrugged. "Fair enough."

"Anyway, we didn't explicitly write 100 in our list, but we can show that it's an element like—"

"Like this, right?" Yuri said, her hand darting forward to write a 0 after my 10.

$$100 \in \{2, 4, 6, 8, 10, 12, 14, 16, \ldots\}$$

"Right. We can use another symbol to show that, for example, 3 *isn't* an element of the set."

$$3 \notin \{2, 4, 6, 8, 10, 12, 14, 16, \ldots\}$$

"Easy peasy. And it doesn't include 1, or 5, or 7—"

"Hold up there. Don't use the word 'include.'"

"Huh?"

"You should say '1 doesn't *belong to* the set,' not 'the set doesn't include 1.'"

"You're even pickier today than usual."

"But it's because—"

"Enough. I'm tired. And hungry."

"You sure do love your snacks," I said, shaking my head.

"Snacks rock. Besides, your mother is calling out to me, I can feel it."

Yuri stood and put a hand to her forehead.

"She seems to be pulling me in this direction..."

Yuri went out my door, heading for the kitchen. I rolled my eyes, but even as I did, my stomach gave a sympathetic growl. I sighed, scooped up the notebook and some pencils, and headed after her.

3.1.3   The Empty Set

By the time I reached the dining room, Yuri was already digging in to a piece of cake.

"You have *got* to have some of this. Wonderful," she said, licking her lips.

My mother appeared from the kitchen, plate in hand.

"Of course he'll have some, won't you honey," she said, placing it on the table.

"I thought you wanted me to teach you about sets," I said, sitting down.

"Yeah, *teach* me. Not give me a million new words to learn."

"Okay, okay. Let's start with some problems, then. That would be more fun, right?"

"You know me all too well."

"First problem, then," I said, opening the notebook. "Is this a set?"

$$\{\}$$

"With no stuff inside?"

"Stuff?" I said. "What's 'stuff'?"

"You know what I mean. The *elements*. So really your question is, does a set have to have at least one element? And my answer is, you tell me."

"And my answer to that is no, it doesn't. A set with no elements is called the empty set."

"Now there's an easy hobby to maintain—a collection of nothing."

"How about this one?" I asked, continuing to write.

$$\{1\}$$

"Sure, that's a set. One that only includes—whoops...I mean, one that has just 1 as an element."

Yuri picked up a pencil.

"So I can write this, right?"

$$1 \in \{1\}$$

"Good. A fast learner, as always. How about this one? Do you think this is true?"

$$\{1\} \in \{1\} \quad ?$$

"Mmm...I guess?"

"Why?"

"Uh...just because."

"Actually, this is not a true statement. This is."

$$\{1\} \notin \{1\}$$

"Huh."

"So 1 is an element of $\{1\}$, but $\{1\}$ is not an element of $\{1\}$. It's important to remember that the number 1 and the set $\{1\}$ are different things."

"Actually, I guess that makes sense. Because you put elements in sets, and a set isn't an element."

"Well..." I said, "not quite."

"Oh, brother."

3.1.4 Sets of Sets

"Take this for example," I said.

$$\{1\} \in \{\{1\}, \{2\}, \{3\}\}$$

"So many braces..." Yuri said, tracing along what I'd written with her fork. "This is a true statement?"

"It is." I brushed a crumb off the page. "Here, let me rewrite the right side with bigger braces on the outside to make it easier to read."

$$\Big\{\{1\}, \{2\}, \{3\}\Big\}$$

"See how this set has three elements, $\{1\}$, $\{2\}$, and $\{3\}$?" I said.

"Ah, okay. So $\{1\}$ here is a set, but it's also an element of a bigger set, right?"

"Exactly. And now that you've realized that, it's easy to see how this is a true statement."

$$\{1\} \in \Big\{\{1\}, \{2\}, \{3\}\Big\}$$

"Okay, finally things are getting interesting," Yuri said, nodding. "It's like you're serving up numbers on plates. Like this."

1	1
2	2
3	3
$\{1\}$	1 on a plate
$\{2\}$	2 on a plate
$\{3\}$	3 on a plate
$\Big\{\{1\}, \{2\}, \{3\}\Big\}$	1 on a plate, and 2 on a plate, and 3 on a plate, all on a tray

"Nice visualization!" I said.

"So I can say this, right?"

$$1 \notin \Big\{\{1\}, \{2\}, \{3\}\Big\}$$

"Sure, because that's a 1 that's not on a plate."

"And I can say this too, yeah?"

$$\{1, 2, 3\} \in \Big\{\{1, 2, 3\}\Big\}$$

"Yep. That's $1, 2, 3$ on a plate, and that's on the tray. I'd say you've nailed this."

Yuri shot me a thumbs-up.

"Try this, then," I said. "Create a set that has as its elements both the number 1 and a set containing 1."

"No problem. Like this, right?"

$$\left\{1, \{1\}\right\}$$

"Perfect," I said.

"Oh," Yuri said, snapping her fingers, "These are both true, right?"

$$1 \in \left\{1, \{1\}\right\} \quad \text{and} \quad \{1\} \in \left\{1, \{1\}\right\}$$

"Sure."

"And those are the only two statements I can write in the form something $\in \left\{1, \{1\}\right\}$? ... Mmm, I guess that's kind of obvious."

"No, that's great! 'Obvious' doesn't mean 'not important.' It's vital that you create examples like this, even the obvious stuff. Good for you for doing it."

3.1.5 Intersections

"Moving on," I said. "Let's talk about using two sets to create a new one."

"Ooh, that sounds fun," said Yuri.

"First, we get a new symbol—one that indicates we want the intersection of two sets. It looks like this."

$$\cap$$

I continued. "Combining two sets with this intersection operator gives you a new set, one consisting of all elements belonging to both the original sets."

"A new set consisting of the ... Wait, I'm lost. Write that down."

"Sure. It looks something like this."

$$\{1, 2, 3, 4, 5\} \cap \{3, 4, 5, 6, 7\} = \{3, 4, 5\}$$

"Let's see," Yuri said, tracing her finger along the equation. "Oh, I get it. This gives you a set of numbers that are in both the first two sets. These here."

Yuri underlined the common elements.

$$\{1, 2, \underline{3}, \underline{4}, \underline{5}\} \cap \{\underline{3}, \underline{4}, \underline{5}, 6, 7\} = \{\underline{3}, \underline{4}, \underline{5}\}$$

"Okay, I'm good," she said.

"Great. This new set is called the 'intersection' of the original ones. We can represent this as a picture using something called a Venn diagram, like this."

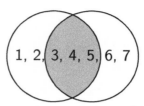

An intersection represented as a Venn diagram

Intersection

The intersection of sets A and B is the set of all elements belonging to both A and B, written as

$$A \cap B =$$
{those elements belonging both to A and to B}.

"I like the picture. Way easier to see what's going on."

"So can you tell me what would be the intersection of these two sets?"

$$\{2, 4, 6, 8, 10, 12, \ldots\} \cap \{3, 6, 9, 12, 15, \ldots\} = ?$$

"Mmm...I guess it would be 6 and 12 and so on?"

$$\{2, 4, 6, 8, 10, 12, \ldots\} \cap \{3, 6, 9, 12, 15, \ldots\} = \{6, 12, \ldots\}$$

"Right! The first set is the set of all multiples of 2, and the second one is the set of all multiples of 3. So what would be a good way to describe their intersection?"

"Multiples of 6? Er, 'the set of all multiples of 6,' to be wordy about it."

$$\{2, 4, 6, 8, 10, 12, \ldots\} \qquad \text{the set of all multiples of 2}$$
$$\{3, 6, 9, 12, 15, \ldots\} \qquad \text{the set of all multiples of 3}$$
$$\{6, 12, \ldots\} \qquad \text{the set of all multiples of 6}$$

"Right! And you get multiples of 6 because that's the least common multiple of 2 and 3."

"Makes sense, I guess, since elements in the intersection have to be both a multiple of 2 and a multiple of 3."

"How about this one?"

$$\{2, 4, 6, 8, 10, 12, \ldots\} \cap \{1, 3, 5, 7, 9, 11, 13, \ldots\} = ?$$

"Let's see. So we're looking for the intersection of . . . sets of even numbers and odd numbers? There's nothing in common, is there?"

"You're right. Remember what we call a set that doesn't have any elements?"

"Oh, right! The empty set!"

$$\{2, 4, 6, 8, 10, 12, \ldots\} \cap \{1, 3, 5, 7, 9, 11, 13, \ldots\} = \{\}$$

"Well done."

3.1.6 Unions

"Let's talk about unions next," I said. "The symbol looks like the one for intersection, turned upside-down. If it helps you keep them straight, think of this one as looking like a 'U' for 'union,' and the other one as looking like an 'n' for 'n-tersection.'"

$$\{1, 2, 3, 4, 5\} \cup \{3, 4, 5, 6, 7\} = \{1, 2, 3, 4, 5, 6, 7\}$$

"I'm pretty sure you can guess what that means."

Yuri nodded. "You're collecting elements from both the sets."

"Yep. A Venn diagram would look something like this."

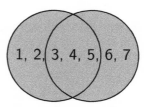

A union represented as a Venn diagram

Union

The union of sets A and B is the set of all elements belonging to at least one of A or B.

$$A \cup B =$$
{those elements belonging either to A or to B}.

Yuri pointed at the equation.

"Some of the elements are in both. Wouldn't you show that by writing it like this?"

$$\{1, 2, 3, 4, 5\} \cup \{3, 4, 5, 6, 7\} = \{1, 2, 3, 3, 4, 4, 5, 5, 6, 7\} \quad ?$$

"Normally you wouldn't, because $\{1, 2, 3, 3, 4, 4, 5, 5, 6, 7\}$ and $\{1, 2, 3, 4, 5, 6, 7\}$ are considered to be the same thing, as sets."

"How can these be the same thing? They're different!"

"Well, yes and no. Sets are formed just according to *what kind* of element belongs to them, not how many. The $\in$ operator only tells you that an element belongs to a set, not the number of that kind of element in the set. That means there's no good way to distinguish between $\{1, 2, 3, 3, 4, 4, 5, 5, 6, 7\}$ and $\{1, 2, 3, 4, 5, 6, 7\}$."

"Huh. Okay.."

"Oh, that's also why the order you list things in isn't important. So for example $\{1, 2, 3\}$ and $\{3, 1, 2\}$ are considered to be the same set."

"Makes sense, in a mathy sort of way."

"Anyway, back to unions. Can you see what this would give you?"

$$\{2, 4, 6, 8, \ldots\} \cup \{1, 3, 5, 7, \ldots\} = ?$$

"Uh... All the evens and odds."

"I think there's a name for that."

"Oh, yeah. The natural numbers."

"Right."

$$\{2, 4, 6, 8, \ldots\} \cup \{1, 3, 5, 7, \ldots\} = \text{the set of all natural numbers}$$

3.1.7   Inclusion Relations

My mother entered the room, carrying a tray of hot drinks that smelled like damp weeds.

"Ugh," I muttered. "Herb tea."

"Did you say something?" she said, frowning.

"Not a thing."

"Herb tea is good *and* good for you. Drink up."

"I don't think I inherited your taste buds," I said.

"It smells wonderful! Thank you!" Yuri chirped.

"Always such a sweet child," Mom said, heading back to the kitchen.

"Anyway, back to sets." I said. "So we have an operation for finding intersections, and one for unions. Both of these take two sets and use them to create a new set."

"Right."

"Okay, a new symbol then. It looks like this."

$$\subset$$

"This one's a little bit different from the others," I continued. "It doesn't create a new set, it shows an inclusion relation between sets."

"What's an inclusion relation?"

"It means one set is a subset of another one."

"Clear as mud. Try harder."

"I think an example would do. Say you've got these two sets."

$$\{1, 2\} \quad \text{and} \quad \{1, 2, 3\}$$

"Okay."

"See how all the elements of the first set are also in the second one?"

"The 1 and the 2. Sure."

"That means the first set is a subset of the other one. You show that like this."

$$\{1,2\} \subset \{1,2,3\}$$

"Ah, okay. Got it."

"You can read this as either 'set $\{1,2\}$ is a subset of set $\{1,2,3\}$,' or 'set $\{1,2\}$ is included in set $\{1,2,3\}$.' This is why I don't want you to say 'include' when you're talking about the *elements* in a set. It's important not to confuse relationships between elements and sets with relationships between two sets. Let me give you some examples."

$$1 \in \{1,2,3\} \qquad 1 \text{ is an element of } \{1,2,3\}$$

$$2 \in \{1,2,3\} \qquad 2 \text{ is an element of } \{1,2,3\}$$

$$3 \in \{1,2,3\} \qquad 3 \text{ is an element of } \{1,2,3\}$$

$$\{\} \subset \{1,2,3\} \qquad \{\} \text{ is included in } \{1,2,3\}$$

$$\{1\} \subset \{1,2,3\} \qquad \{1\} \text{ is included in } \{1,2,3\}$$

$$\{1,2\} \subset \{1,2,3\} \qquad \{1,2\} \text{ is included in } \{1,2,3\}$$

$$\{1,2,3\} \subset \{1,2,3\} \qquad \{1,2,3\} \text{ is included in } \{1,2,3\}$$

"Wait, the empty set is a subset of $\{1,2,3\}$?"

"Sure."

"And $\{1,2,3\}$ is a subset of itself?"

"Yeah. You can also write that like this, for clarity."

$$\{1,2,3\} \subseteq \{1,2,3\}$$

"You can use the normal subset symbol along with this under-lined one to distinguish between a 'proper' subset—one that doesn't include all the elements of the base set—and cases where a set is a subset of itself. Just be sure to define things so that it's clear what you're doing."

"Okay, okay... By the way, $\{2\}$ works too, right?"

"What do you mean, 'works'?"

"That $\{2\}$ is part of $\{1, 2, 3\}$."

"C'mon Yuri. Use the right words. Say that $\{2\}$ *is included in* $\{1, 2, 3\}$, or that $\{2\}$ *is a subset of* $\{1, 2, 3\}$. Not that it's a 'part' of it."

"But anyway, I *can* say that $\{2\}$ is a subset of $\{1, 2, 3\}$, right?

"Yes, you can."

3.1.8   Why are Sets Important?

Yuri called for a break and we went back to my room. She headed straight for the jar of candy I kept on my desk.

"I guess it was sorta fun, learning a bunch of new symbols and all," she said, fishing around in the jar. "But are sets really such a big deal?"

"Sure. Sets make it lots easier to organize some mathematical concepts. You'll see these symbols pop up in math books all the time."

Yuri pulled a lemon-flavored candy out of the jar and popped it into her mouth.

"But *why*?" Yuri turned and gave me a deadly serious look. "Unions and intersections and all that stuff seems kind of obvious. What's all the fuss about?"

"I'm not sure I can give you a good answer. I'll ask Miruka."

"Miruka! Yeah, I'll bet she knows! I wish I could ask her myself..."

"Come to my school and you can."

"Your school? By the time I'm in your school, you guys will already be graduated!"

"Huh?"

Yuri's comment made me realize that Miruka and I only had a little over a year left in high school. It was hard to imagine.

"You should invite her over here! Maybe if you tell her you've got some yummy chocolate..." she sang.

"That seems like kind of a weak lure."

"I don't care how you do it, just get her over here!"

I didn't respond. My mind had already drifted back to math.

Just why are *sets so important?*

3.2 ADDING LOGIC

3.2.1 Intensional Definitions

Miruka answered my question the following Monday.

"So we can deal with infinities," she said.

"Infinities?" I replied. "Why are sets needed to create infinities?"

We were in the library, me in my usual seat and Miruka standing with her back to the window. The late afternoon sun ignited a nimbus around her elegant form, making me wince. We were alone that day, which was somehow relaxing.

"Deal with them, I said, not create them," Miruka said. "That's one reason they're important, at least."

"Hang on, there's plenty of finite sets out there."

"Boring ones, sure. But sets really come into their own when they capture the infinite. Without sets and logic, infinities are tricky things to work with."

"Sets ... and logic?"

I paused to think. I'd be the last person on earth to say logic wasn't important, but to me, sets and logic seemed like two completely different things. On the one hand you have a collection of elements, on the other a set of rules for making sure you're doing mathematical proofs correctly.

Miruka chuckled—probably at my expression—and began pacing between our table and the window. She whirled about at each reversal of direction, her hair cascading out in soft waves as she did.

"Sets are all about belonging or exclusion," she said. "With logic it's true or false. Both present two contrasting alternatives. If you can move beyond extensional definitions of sets, and starting thinking in terms of intensional definitions, their relationship is clear."

Miruka's eyes took on that glimmer that indicated she was slipping into full-on math mode. I braced myself to get schooled in set theory.

I started to ask a question, but Miruka had anticipated it and already begun her explanation.

"An extensional definition is when you define a set by lining up a bunch of elements," Miruka said, snatching my notebook and pencil.

"That's what you used to teach Yuri, I assume?"

$$\{2, 4, 6, 8, 10, 12, \ldots\} \quad \text{an extensional definition}$$

I nodded, and Miruka resumed her pacing.

"They use concrete values, so they're easy to understand. But they're a limited approach when dealing with infinitely large sets, since you obviously can't list infinitely many elements. Those three dots are convenient, but they're also vague and limited. They can't always tell the whole tale."

"And the other kind of definition avoids that problem?"

Miruka nodded.

"Intensional definitions define sets using propositions, require-ments to be fulfilled for membership. They use *logic* to define sets. In this case we could have used the proposition 'n is a multiple of 2.' Then we write the definition, separating the name of the element and its proposition with a vertical bar."

She paused her pacing long enough to write another example in the notebook.

$$\{n \mid n \text{ is a multiple of } 2\} \quad \text{an intensional definition}$$

I nodded. "That certainly gets rid of the vagueness of ellipses."

"Which makes the requirements for set membership perfectly clear, avoiding misunderstandings even when there are infinitely many elements in the set. So when you're dealing with the infinite, intensional definitions are far easier to deal with than extensional definitions."

Miruka slid into the seat next to me.

"Intensional definitions are useful," she said, "but there are a cou-ple of things to watch out for. For one, there's often more than one way to define the same set. For example, all of these are equivalent sets."

$$\{n \mid n \text{ is a multiple of } 2\}$$
$$\{x \mid x \text{ is a multiple of } 2\}$$
$$\{n \mid n \text{ is even}\}$$
$$\{2n \mid n \text{ is a natural number}\}$$

"But more importantly, abusing them can lead to contradictions."

"Contradictions? You mean the proposition used to define them can be both true and false?"

"Let me show you a famous example," she said, leaning in close to whisper in my ear. "Russell's paradox."

3.2.2   Russell's Paradox

Miruka straightened up and grinned.

"Russell's paradox is an example of how contradictions arise if you insist you can use any proposition at all to define sets. Here's how it works, using the proposition $x \notin x$."

Problem 3-1 (Russell's paradox)

Show how defining the set $\{x \mid x \notin x\}$ leads to a contradiction

---

"In the interest of time, how about you walk me through this one," I said.

"I'd love to," Miruka replied. "Start off using the proposition $x \notin x$ to define a set R."

$$R = \{x \mid x \notin x\}$$

"A simple question reveals the paradox: Is R an element of itself? In other words, is the set $\{x \mid x \notin x\}$ a member of the set $\{x \mid x \notin x\}$? We should be able to get a simple yes–no answer, right?"

"Yeah, sure. Like you said, logic is all about true or false. There's not really anything in-between."

"And how would you test for truth or falsity in this case?"

"Well, you'd just see if assuming that $R \in \{x \mid x \notin x\}$ led to a logical contradiction. If it does, then R isn't an element of R."

"Have at it, then," Miruka said, smiling.

"Okay," I said, growing wary. "So assume this is a true statement." I wrote in the notebook.

$$R \in \{x \mid x \notin x\}$$

"That would mean R is an element of the set $\{x \mid x \notin x\}$. In other words, R isn't an element of R."

$$R \notin R$$

"Then we can replace the R on the right with the proposition, and we get this."

$$R \notin \{x \mid x \notin x\}$$

"But that contradicts our original assumption, because it says R is *not* an element of the set $\{x \mid x \notin x\}$, so $R \in \{x \mid x \notin x\}$ is a *false* statement, and $R \notin \{x \mid x \notin x\}$ must be true."

"Maybe you should walk through that one, too," Miruka said, winking. "Just to be sure."

"Fair enough," I said. "So now we assume that this is true."

$$R \notin \{x \mid x \notin x\}$$

"In other words, R must be an element of R."

$$R \in R$$

"Like before, we can replace the R on the right with the proposition, giving us this."

$$R \in \{x \mid x \notin x\}$$

"This tells us that . . . oh. It tells us that R *is* an element in the set, so we have a contradiction again."

"Exactly," Miruka said. "So the proposition $R \in \{x \mid x \notin x\}$ cannot be shown to be either true or false. Therein lies our paradox."

Answer 3-1 (Russell's paradox)

When we investigate whether the set $\{x \mid x \notin x\}$ is an element of itself, we obtain contradictions both when we assume it is an element and when we assume it is not.

Miruka leaned back, her hands behind her head.

"There's no quick-and-easy fix for Russell's paradox. After all, it's a problem associated with the $\in$ operator, one of the most important tools for working with sets. The best we can do is limit what kind of intensional definitions are allowed in our definitions."

"What kind of limit?" I asked.

"Like setting up a universal set U and using it as the pool you fish for elements from. So instead of using a naked predicate $P(x)$ as your proposition like this...

$$\{x \mid P(x)\}$$

... you ensure that the only elements you'll deal with are those taken from set U, like this."

$$\{x \mid x \in U \overset{\text{and}}{\wedge} P(x)\}$$

"That will make your intensional definitions safe."

3.2.3 Set and Logic Operators

Miruka continued with her lecture on sets.

"When you create intensional definitions, you're using propositions to define your sets. So it isn't so strange that there's a relationship between sets and logic. There's even a clear correspondence between set and logic operations."

Sets	$\longleftrightarrow$	Logic
Set $A = \{x \mid P\}$	$\longleftrightarrow$	Proposition P
Set $B = \{x \mid Q\}$	$\longleftrightarrow$	Proposition Q
Intersection $A \cap B$	$\longleftrightarrow$	Logical AND, $P \wedge Q$ (P and Q)
Union $A \cup B$	$\longleftrightarrow$	Logical OR, $P \vee Q$ (P or Q)
Universal set U	$\longleftrightarrow$	True
Empty set	$\longleftrightarrow$	False
Complement $\overline{A}$	$\longleftrightarrow$	Negation $\neg P$ (not P)

"It's been a while since I've looked at sets this deeply," I said. "What's this complement $\overline{A}$ again?"

"The set of elements in the universal set U that do not belong to A."

"Right, right."

"De Morgan's laws are particularly beautiful."

Sets	$\longleftrightarrow$	Logic
$\overline{A \cap B} = \overline{A} \cup \overline{B}$	$\longleftrightarrow$	$\neg(P \wedge Q) = \neg P \vee \neg Q$
$\overline{A \cup B} = \overline{A} \cap \overline{B}$	$\longleftrightarrow$	$\neg(P \vee Q) = \neg P \wedge \neg Q$

"Beautiful how?"

"Each of the four expressions here can be represented by a single pattern. Do you see it?"

I thought for a moment, but I ended up shaking my head.

"I don't think so, no."

"Take a look at this," Miruka said.

$$h(f(x,y)) = g(h(x), h(y))$$

"There are three functions here, $f(x,y)$, $g(x,y)$, $h(x)$. Write them out explicitly like this, and you can get all of De Morgan's laws."

$f(x,y)$	$g(x,y)$	$h(x)$	$h(f(x,y))$	$=$	$g(h(x), h(y))$
$x \cap y$	$x \cup y$	$\overline{x}$	$\overline{A \cap B}$	$=$	$\overline{A} \cup \overline{B}$
$x \cup y$	$x \cap y$	$\overline{x}$	$\overline{A \cup B}$	$=$	$\overline{A} \cap \overline{B}$
$x \wedge y$	$x \vee y$	$\neg x$	$\neg(P \wedge Q)$	$=$	$\neg P \vee \neg Q$
$x \vee y$	$x \wedge y$	$\neg x$	$\neg(P \vee Q)$	$=$	$\neg P \wedge \neg Q$

Miruka continued. "Using logic to define sets with intensional definitions sounds like an abstract concept, but representing them with logic allows us to study them mathematically. Sure, we have to be careful not to stumble into contradictions, but the payoff is huge. You can use sets and logic to describe topics from abstract algebra, geometry, analysis... the list goes on. But that's not all. Sets allow us to mathematically study *mathematics itself*."

I was reeling from Miruka's impassioned paean to set theory, but this last statement brought me back down to earth.

"The mathematical study of mathematics? How does that work?"

Before Miruka could answer, we heard a loud crash followed quickly by a high-pitched yelp.

I turned toward the library entrance to see Tetra, bent over and rubbing her knee.

3.3 INFINITY

3.3.1 *Bijections and Birdcages*

"Wow, that stings," Tetra said, still rubbing her knee.

"What happened?" I asked.

"This book cart happened. Did Ms. Mizutani leave this here? Somebody could get seriously hurt."

"That cart's been there for as long as I can remember," I said.

"Yeah? Sorry, I get confused when there's too much stuff in a room."

Tetra limped toward us.

"Maybe you should focus on what's right in front of you. Especially while you're walking."

"Sounds like a good strategy."

Tetra pulled up a chair and sat with us.

"So what's on the menu for today?" she asked.

I gave her a quick outline of what we'd discussed about sets, logic, and infinity.

"Infinity's so weird," Tetra said. "Just imagine, something that can never be counted, no matter how fast you go or how long you try."

Miruka raised a finger.

"It's probably even weirder than you're imagining. For example, did you know that some infinities are bigger than others?"

"Huh? How can something be bigger than infinity?"

"Before I answer that, a little demonstration."

Miruka held her hands out, one over the other, palms facing. As she spoke, she touched the fingers of each hand together, thumb to thumb, index to index, middle finger to middle finger . . .

"Even if we don't count two groups of things, we can match them up one-by-one to compare the number of things in each group."

Each finger was now paired with its partner, turning Miruka's hand into something like a birdcage.

"Even if I didn't know how many fingers were on either hand, I could now at least claim that each hand had the same number of fingers, right?"

"Sure, I guess," Tetra said.

"And we could do something similar with any two sets of finite objects, couldn't we?"

"Hmm ... That seems like quite a jump from just five pairs of fingers."

"Let me convince you, then."

Miruka turned to a new page in the notebook, and sketched a diagram.

"Say these are our sets, and we've found a way to associate each element in one set with a unique element in the other. When we're done, there are no unmatched elements in either set. A mapping like this is called a bijection, by the way."

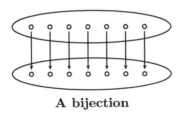

A bijection

Tetra raised her hand.

"Question. What's a mapping?"

"Think of it as a way of linking things together," I said. "Like how she touched the fingers on each hand. Thumb goes to thumb, pinky goes to pinky, and so on. It's just a way of associating the things in one set with those in another."

"Oh, stop being so wishy-washy about it," Miruka said, scowling. "Tetra, say you have two sets, A and B. For each element in set A, you create an association with a single element in set B. That association is called a mapping from set A to set B. A function, in other words, just not necessarily one where the inputs and outputs are real numbers."

Miruka raised an eyebrow at me in a see-how-it's-done? look, then continued.

"There are different kinds of mappings," she said. "But for now we're interested in just three—surjective, injective, and bijective."

She turned back to the notebook and began sketching more graphs.

"A surjective mapping is one where every element in the second set gets something mapped to it. It's okay if more than one element in the first set goes to the same element in the second."

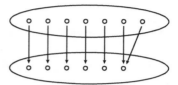

A surjective mapping maps to all elements

"If you miss anything, the mapping isn't surjective."

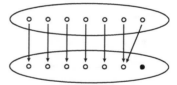

A non-surjective mapping misses some elements

"If everything in the first set gets mapped to something different, the mapping is injective. It's okay if you miss some elements in the second set."

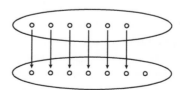

An injective mapping creates unique correspondences

"If two elements in the first set map to the same element in the second, the mapping isn't injective."

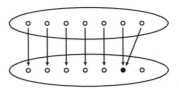

Two elements map to the same element, so this mapping isn't injective

"If a mapping is both injective and surjective, then it's bijective. In other words, there's a one-to-one correspondence between elements in the two sets."

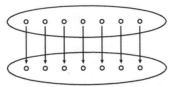

A bijective mapping creates a one-to-one correspondence

"If you have a bijective mapping, you can also create an inverse mapping."

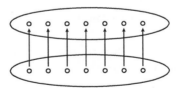

Bijective mappings can produce inverse mappings

Miruka put down the pencil and looked at Tetra.

"So do you see how a bijective mapping lets us claim that both sets have the same number of elements?"

"Sure, that makes sense," Tetra said, creating her own bijective birdcage and peering into it.

"So let's expand our horizons, and apply mappings like this to infinite sets." Miruka's voice dropped to a whisper. "But I warn you—when we follow these mappings into the infinite, we wander some weird paths. So weird, in fact, that even Galileo turned back."

My ears perked up.

"Did you say Galileo?"

3.3.2 New Concepts

"Yes, *the* Galileo," Miruka said. "He starting having doubts about all this when he noticed you could create a bijection between the natural numbers and squares, like this."

1	2	3	4	5	6		n	All natural numbers
$\updownarrow$	$\updownarrow$	$\updownarrow$	$\updownarrow$	$\updownarrow$	$\updownarrow$	$\cdots$	$\updownarrow$ $\cdots$	
1	4	9	16	25	36		n^2	All square numbers

"What does this imply?" Miruka asked, tapping the notebook and looking at Tetra.

"Well, we were just saying that if you have a bijection between two sets, that means both sets have the same number of elements, right?"

Miruka gave an impatient nod. "And what does *that* imply?"

Tetra's face clouded, and she paused before speaking again.

"Well... it seems like that would mean there are just as many square numbers as there are natural numbers. But that doesn't make any sense!"

"Galileo would agree with you. After all, the squares are just one part of all the natural numbers."

$$①, 2, 3, ④, 5, 6, 7, 8, ⑨, 10, 11, 12, \ldots$$

"He didn't think it made sense to say that a whole and a part of that whole can have the same number of elements. So in the seventeenth century, Galileo claimed that you just can't apply bijections to infinite sets, and turned away from exploring this path."

Miruka leaned forward to add a line to our notebook.

Galileo

> You can't apply bijections to claim that infinite sets have equal size.

"It wasn't until the nineteenth century that mathematicians like Cantor and Dedekind took up exactly the same mathematical facts but looked at them in a new way. Dedekind turned the view of Galileo on its head, claiming that the existence of a bijection between the whole of a set and one of its subsets is *the very definition of infinity*."

Dedekind

> An infinite set is one for which there exists a bijection between itself and one of its proper subsets.

"Cantor went even deeper, studying the 'number' of elements in an infinite set, which we now call its cardinality."

Miruka paused and scanned our faces before continuing.

"When you think you've found an error, it's no mistake to consider your current actions a failure and retreat. But what makes Dedekind a genius is that he turned failure into discovery. He took what looked to be a mistake, what looked like nonsense, and transformed it into something amazing. Sure, sometimes pulling back is the proper strategy. But keep in mind that what you've run into may actually be the boundary between what is currently known and a whole new concept, a boundary that's begging to be broken through.

"When they say there's no number you can add to 1 to get 0, tell them there is, and call it -1.

"When they say there's no number whose square is 2, tell them there is, and call it $\sqrt{2}$.

"When they say there's no number whose square is -1, tell them there is, and call it i.

"When they say you can't create a bijection between a set and one of its subsets, tell them you can and call it the definition of an infinite set.

"Extending existing concepts never comes easy. Genius lies in realizing that that difficulty is the necessary pause before you can leap into flight."

I swallowed and nodded, and Tetra whispered, "Wow."

"It's easy to see how much doubt all these new concepts caused," Miruka said. "Just look at their names."

"How do you mean?" Tetra asked.

"What do you call a number like -1?"

"A negative number."

"What about a number like $\sqrt{2}$?"

"An irrational number."

"And i?"

"An imaginary number."

"Negative, irrational, imaginary..." Miruka said, standing from her seat. "Don't those names give a clue as to how their discoverers felt about what they'd found?"

She walked to the window, where she fell silent and looked outside. The fervor with which she'd been discussing math had vanished.

"Everyone has their doubts when faced with a new path to tread."

3.4   DESCRIPTIONS AND EXPRESSIONS

3.4.1   Heading Home

Miruka mumbled something about piano practice with Ay-Ay and left without so much as a good-bye. Tetra and I had had our fill of studying for the day, so we decided to head home. We made our way along the winding streets toward the train station.

I was going through everything Miruka had talked about, summarizing everything out loud, but mostly to myself.

"Sets and logic... In an intensional definition, it's logic that defines the set. Anything that fulfills the requirements of a proposition is taken to be an element in the set. It's like expressing things as propositions results in the formation of this phenomenon called a 'set.' The conditions for the thing result in the thing itself."

"Maybe it works something like this," Tetra said, choosing each word with care. "When we say we can describe something mathematically, in essence we're talking about how we can write it out. It's impossible to actually write out infinitely many objects, but what we *can* do is describe some characteristic that those objects have in common."

I continued walking in silence. Tetra was clearly just as lost in her own thoughts as I was in mine.

"I guess it's no coincidence that the word 'describe' contains the Latin root for 'to write.' Maybe being able to write something down is a prerequisite for being able to describe it? It's interesting how different that word is from 'express,' even though they mean nearly the same thing. With 'de-scribe' you're writing something down. With 'ex-press' you're pushing it out. Hmm... Well what about 'represent,' then? And 'denote'?"

Tetra stopped, rummaged through her knapsack, and pulled out a thick book.

"What's that?" I asked. "A dictionary?"

Her gaze jumped to me, as if surprised I'd spoken.

"Oh, sorry. Ignore me, I was just thinking about something."

"I know. You were describing, expressing, and representing your thoughts out loud again."

3.4.2   At the Bookstore

Tetra said she wanted some help picking out books, so we stopped by the bookstore. We headed for the math section and stood there for a time, browsing the shelves.

"I dunno," Tetra said, "maybe I shouldn't be buying another book after all. I buy too many as it is already."

"Still looking for one that only uses notation you're used to, so you don't fret over the differences?" I joked.

"Nah, I'm over that, I think. It's just that when I start studying something, I get this weird urge to buy lots of books about it. As if somewhere inside one of them is the secret to gaining the insight that everyone else seems to have picked up on. Like reading a spoiler in a game strategy guide."

"There's no final boss in math, you know," I laughed.

"Stop teasing," Tetra said, laughing along with me. "And yeah, I know. With math it's not about having the right book, or even *reading* the right book. It's all about effort, thinking hard and putting pencil to paper, as you say. Still, it's hard to shake the feeling that having *that one perfect book* would be a big help."

Tetra took one of the books from the shelf, flipped through a few pages, then replaced it.

"Actually, I know what you mean," I said. "Problem is, 'that one perfect book' is a moving target."

"What do you mean?"

"Well, when it comes to math, I think everyone gets stuck on different things at different times for different reasons. Some things will come easily, others ... not so much. Sometimes a flash of insight about some key point will push you way ahead. So if you're going to rely on a book, you have to be sure you understand exactly what it is you're having problems with, and find one that addresses that specific need."

"Whoa, that's good," Tetra said, taking a step closer. "Keep going, with specifics."

"Specifics? Well, uh ... okay, say you're having problems with mathematical induction. That's not really enough to go on, if you want to make progress. You need to look in the mirror and ask yourself exactly what it is about induction that's unclear. And no cheating and saying 'all of it!', either. You need to look deep inside yourself, go searching for your border between understanding and confusion. *That's* when you turn to books, and look for the pages that are right there, straddling that border. Then you start reading, and thinking, and asking yourself if what's in that book can answer your questions. You should look at other books, too, to see what new perspectives they can bring. Keep doing that, and you're sure to find 'that one perfect book' eventually. The one that's perfect for you at that particular time, at least."

"Wow, sounds like that would take a long time."

"It does. Math is a marathon, not a sprint."

"I think I understand it best when you're the one teaching me," Tetra said, looking down. "Too bad I can't keep *you* on my bookshelf at home."

3.5 NULL ANSWER

"So sets are all about making it easier to deal with infinity?" Yuri asked, scrunching her face. I had just finished giving her a rundown of Miruka's latest lecture.

"Is infinity really all that big a deal?" she continued. "I don't get it."

"Honestly, I don't get infinity either," I said. "I'm working on it, though."

"Well don't work on it too hard. I already feel like I'm infinitely far behind you guys. I want to catch up!"

"No need to rush. The math isn't going anywhere. Besides, you're farther ahead than you think. You think hard, you're able to put into words exactly what you don't understand. Those are some of the most important math skills you can have."

"You sound like a math teacher trying to convince someone they aren't an idiot."

"No, seriously. Keep going the way you are, and some day you'll be teaching Miruka a thing or two."

"Hah! Yuri's introduction to set theory."

Yuri took off her glasses and put them aside.

"Here's a problem for you," she said, leaning towards me. "Will I ever be included in the set of beautiful girls?"

"You can't create an inclusion proposition based on a subjective opinion."

"I can if I use you as my beauty detector."

"I ... uh ..."

"C'mon, out with it. Let the set of all girls you personally know be the universal set. Am I an element of the subset of pretty ones?"

"There's only one safe answer to that problem," I said. "Neither yes nor no, just the null set."

When a mathematician wants to
introduce some new concept, all it
takes is a clear definition to cause the
abstract to materialize from the Void
and descend to earth as a set and its
member elements, who are then free to
mix and mingle with all the other
mathematical objects that preceded
them.

KOJI SHIGA
A Flight Into Infinity

CHAPTER 4

Know Your Limits

> "Now Cinderella, depart; but remember, if you stay one instant after midnight, your carriage will become a pumpkin, your coachman a rat, your horses mice, and your footmen lizards; while you yourself will be the little cinder-wench you were an hour ago."
>
> CHARLES PERRAULT
> *Cinderella*, trans. Unknown

4.1 AT HOME

4.1.1 Yuri

"Arrrgh!" Yuri bellowed, storming into my room one Saturday in February. She threw her bag across the room, where it collided with my bookshelf.

"Whoa, what's up with you?" I asked.

I'd heard Yuri enter my house and say hi to my mom, chipper as usual, so this display was more than a little unexpected.

"I let a boy get the best of me yesterday. Gah, I'm so mad. I *hate* losing."

Yuri shook her head, whipping her ponytail from side to side.

"You got in a fight?"

"No way I'd lose a fight with that dweeb. He beat me in *math*—that's what's so infuriating."

Yuri pulled a notebook out of her bag, turned to a page, and slammed it on my desk.

"This problem," she said.

Problem 4-1
Is the following a true statement? $$0.999\cdots = 1$$

"Ah, a classic," I said. "What was your answer?"

"Something along the lines of 'of course not, moron.'"

"Why'd you say that?"

"Because it *isn't*! I mean, look at it! It's zero-point-lotsa-nines. It's gotta be a little bit less than 1!"

"And what did he say?"

"He got all full of himself, said he could prove it was true."

4.1.2   The Boy's Proof

Yuri turned the page in her notebook to reveal a proof in a handwriting that was not her own.

Answer 4-1

Clearly, 1 equals 1.

$$1 = 1$$

Divide both sides of this equation by 3, writing the left side in decimal and the right side as a fraction.

$$0.333\cdots = \frac{1}{3}$$

Multiply both sides by 3.

$$3 \times 0.333\cdots = 3 \times \frac{1}{3}$$

Calculate both sides.

$$0.999\cdots = 1$$

Thus, $0.999\cdots = 1$.

"Not bad for a middle school student," I said. "The kid has potential."

Yuri jabbed my shoulder.

"You aren't allowed to take his side! Is this right, by the way?"

"The rigor could be improved, but yeah, pretty much."

"Bah!" Yuri slumped into the chair next to my desk. "To be honest, after I got home that day I came up with my own proof. But I was hoping it'd be wrong."

"Why's that?"

"Because I want the equals sign to mean things are absolutely, positively, right-on-the-nose the same! That's what's cool about math, that it can be *right*, no questions asked. I want to say that $0.999\cdots < 1$, none of this 'pretty much equals' garbage."

"How about you show me your proof, and *then* we can talk about how all this is right or wrong."

"I've got a better idea. How about I show you my proof, and you explain how it can't be true?"

"How about we just follow the math, and see where it takes us?"

"Deal!"

4.1.3   Yuri's Proof

"Okay," I said. "Show me your proof."

"My *wrong* proof. You've just gotta show me how."

"Proceed."

"Okay, so I started thinking about how you can start with 0.9, then add a nine to get 0.99, then 0.999 and so on."

"Good so far."

"Like, 0.9 is pretty close to 1, but it's still 0.1 from getting there."

"You're talking about the difference between 1 and 0.9, yeah?"

I wrote an equation in the notebook.

$$1 - 0.9 = 0.1$$

"The difference. Exactly. Yeah, I should have used equations I guess. Here, gimme that pencil."

I handed the pencil to Yuri, and she spun the notebook toward herself.

"So there's 0.9, but then we can do 0.99."

$$1 - 0.99 = 0.01$$

"And so on and so on."

$$
\begin{aligned}
1 - 0.9 &= 0.1 \\
1 - 0.99 &= 0.01 \\
1 - 0.999 &= 0.001 \\
1 - 0.9999 &= 0.0001 \\
1 - 0.99999 &= 0.00001 \\
&\vdots
\end{aligned}
$$

"If you repeat this *infinitely many times*, you end up with the difference with 1 being $0.000\cdots$."

$$1 - 0.999\cdots = 0.000\cdots$$

"On the right, you've got $0.000\cdots$, which is just 0, right?"

$$1 - 0.999\cdots = 0$$

"So if the difference between 1 and $0.999\cdots$ is 0, then $0.999\cdots$ must equal 1!"

$$0.999\cdots = 1$$

"And that's my proof," Yuri said, putting the pencil down.

"That's well thought out," I said. "An excellent job for someone in middle school."

"Thank you, but you really need to cut out that 'for someone in middle school' stuff. It's annoying."

"Sorry. But anyway there's one thing we definitely have to clean up—the part where you talk about doing something 'infinitely many times.' That's kind of mathematically sloppy."

"I figured. Seems like no matter how far you go, that $0.000\cdots$ is just a *liiitle* bit bigger than 0. But that would mean that $0.999\cdots$ is just a *liiitle* bit smaller than 1, so I'm kinda hoping that's the case."

"Hmm... I don't know."

"At least give me a shot at showing you how that might work."

"By all means."

4.1.4   Yuri's Alternate Proof

Yuri turned back to the notebook.

"Okay, so check this out," she said. "Obviously 0.9 is smaller than 1, right?"

$$0.9 < 1$$

"But so is 0.99."

$$0.99 < 1$$

"And so on and so on."

$$
\begin{array}{rcl}
0.9 & < & 1 \\
0.99 & < & 1 \\
0.999 & < & 1 \\
0.9999 & < & 1 \\
0.99999 & < & 1 \\
& \vdots &
\end{array}
$$

"So doesn't that show that 0.999 is less than 1?"

$$0.999\cdots < 1 \quad ?$$

Yuri put down her pencil and shrugged.

"Seems just as right as the first one, anyway. But one of them must be wrong."

"An interesting dilemma," I said. "On the one hand, it seems like we can say 0.9, 0.99, 0.999 ⋯ gets as close to 1 as you like. On the other, it seems you can say that it never gets there."

"Well?" Yuri said, looking up at me with pleading eyes. "Which is it?"

4.1.5 My Explanation

I turned to a new page in the notebook, flattening the crease as I gathered my thoughts.

"Okay," I said. "Let me start with something like what you just said, but written out a little bit different. We'll start with a sequence like this."

$$
\begin{aligned}
a_1 &= 0.9 \\
a_2 &= 0.99 \\
a_3 &= 0.999 \\
a_4 &= 0.9999 \\
a_5 &= 0.99999 \\
a_6 &= 0.999999 \\
&\vdots \\
a_n &= 0.\underbrace{9999\cdots 9}_{n \text{ 9s}} \\
&\vdots
\end{aligned}
$$

Yuri nodded. "So you're naming them all as a with a subscript. Got it."

"Also, the subscript shows how many 9's there are. So that would seem to present this dilemma."

(1) The more 9's there are, the closer a_n is to 1.

(2) No matter how big n is, a_n is less than 1.

"Yes, exactly," Yuri said. "One of those has to be wrong. Right?"

"Well, no. These are both true statements."

"Huh? So $0.999\cdots$ is less than 1 after all? But I thought you said—"

"No, that's not right, either. $0.999\cdots = 1$ is a true statement, too."

"I am *so* confused," Yuri said.

She frowned as she concentrated on what we'd written. I left her alone to give her some time to ponder it all. There are few things more important to doing math than time to think. I thought about changing my slogan to 'Silence is the key to—'

A loud clatter of pans came from the kitchen. Apparently my mother was cooking again.

Yuri perked up with a grin.

"I've got it! We just have to change the definition of 'equals,' so that teeny tiny differences don't matter! Mathematicians change definitions all the time, right?"

It took me a moment to recover from the audacity of her suggestion.

"That's an amazing answer, Yuri. So amazing that I wish it were true. But we can't change what equals means. It really does mean absolutely, positively, right-on-the-nose the same."

Yuri's face fell.

"I give up, then. I don't get it."

Just then, a wail of despair came from the kitchen. Yuri and I ran downstairs to see what had happened.

"What's wrong?" I asked my mother, who was standing in front of the open refrigerator in an apron. She turned to me, pale and wide-eyed.

"We're out of eggs! I'd forgotten that I used them all last night!"

"Seriously? That's *it*?"

"But—but I was going to make omelettes!"

"Can't you make something else?"

"I've already got everything else prepared! I..." Mother's expression softened, and the tone in her voice became cloyingly sweet. "Honey, I know you're studying and all, but I don't suppose—"

"Awww, mom. It's freezing out there!"

"C'mon," Yuri said, laughing, "I'll go with you."

4.2   AT THE SUPERMARKET

4.2.1   Arriving at Your Destination

We rode my bicycle to the supermarket, Yuri standing on the rear hub. I was thankful for the exercise to keep warm—it was even colder than I'd feared.

After a short search we found the eggs, paid at the register, and started back. Just as we were heading for the door, Yuri snagged my arm.

"Ooh, look what they've got!" She was pointing at the in-store snack counter. "Ice cream!"

"We've got to get back, Yuri. Mom's waiting for the eggs. Besides, ice cream in this weather? Are you insane?"

"Ice cream is one of those delicious-all-year-round foods," she said, maneuvering in front of me to cut off my escape route.

"Pleeease?" she said, making sad puppy eyes.

I sighed in resignation. We bought two vanilla cones and sat at the counter to eat them.

"Happy now?" I asked.

"Perfectly," Yuri said with a huge smile.

"At least one of us is," I grumbled. "And at least you aren't raging like you were before."

"I have no idea what you're talking about. Delicate flowers do not 'rage.'"

We sat there, licking our ice cream for a while.

"So what do you want to be when you grow up?" Yuri asked.

"I dunno. How about you?"

"Hmm... A lawyer, maybe."

"Sounds like you've been watching too many crime dramas lately."

"That's—that's possible. It would still be pretty cool, though."

Yuri turned her cone in her hand, checking for wayward rivulets of ice cream. I pulled a mechanical pencil out of my breast pocket and looked around for something to write on. I settled on the back of a nearby store flier.

"Would it bug you if your wife made more money than you?" Yuri asked.

"Huh? Where did that come from?" I said, sketching a graph.

"Just wondering. I guess you wouldn't care. You're too oblivious to what people think."

"Whatever. Here, check this out," I said, pushing the graph toward her. "This is what we were talking about before."

$$\underset{1}{\overset{0.9 \qquad\qquad 0.99 \;\; 0.999}{\vrule width 0pt height 0pt \rule{0pt}{0pt}}}$$

"I know," Yuri said.

"See how this sequence $0.9, 0.99, 0.999$ is getting arbitrarily close to 1? It can get as close as you want it to be, because its destination is $0.999\cdots$."

"I said, *I know*. I understand that the sequence can get as close to 1 as you want it to. I get it. Done."

"So what's the problem?"

"That *closer* ain't *is*. Your explanations still aren't showing me how $0.999\cdots$ and 1 can be the same."

Glowering, Yuri took a joyless lick at her ice cream.

"Okay, how about this. I'm going to ask you some questions, you answer yes or no."

"Fine."

"If you continue the sequence $0.9, 0.99, 0.999$ on and on, will you eventually reach 1?"

"No. It doesn't matter how many nines you add, you still aren't there."

I nodded. "That's correct."

Yuri made a scary growling noise.

"You're *trying* to making me mad, aren't you."

"Hang on, hang on. I'm not done yet. One more question: as you continue the sequence $0.9, 0.99, 0.999$, is there some number that you're getting closer and closer to?"

"Yes, you're getting closer to 1. We've been through this like a million times."

"Right, but let me add one more thing. A notational rule, a way we use symbols to represent an idea."

"Okay, what's the rule?"

"That when we have a sequence like $0.9, 0.99, 0.999$, one that can get arbitrarily close to some number, we represent that 'some number' like this."

$$0.999\cdots$$

Yuri froze mid-lick, her eyes wide, staring at what I'd written.

She held up a hand and said, "Hold up a minute. So this is..."

I waited while she did some internal processing. Harsh winter sunlight gleamed gold in her hair.

"Okay," she said, "I think I finally get this. Let me make sure."

"Absolutely."

"This $0.999\cdots$, it's a representation of *some number*."

"Right."

"And that *some number* that it's representing is the number that it's getting closer and closer to."

"That it gets arbitrarily close to, yes."

"Arbitrarily close to, then. But because it never gets there, that *some number* never shows up."

"That's right."

"And that *some number* that $0.999\cdots$ represents is none other than 1."

"Exactly."

Yuri sighed.

"Okay, not only do I understand this now, I now understand what I didn't understand."

"Namely?"

"That $0.999\cdots$ isn't a number. Not the kind I'm used to, at least. It's a representation of a number."

Yuri jotted down some notes, pausing to catch escaping streams of melting ice cream as she did.

- $0.999\cdots$ represents *some number*.

- Continuing the sequence $0.9, 0.99, 0.999$, we can get arbitrarily close to that *some number*.

- But we never reach that *some number*, so it doesn't show up in the representation.

- The *some number* that $0.999\cdots$ represents is equal to 1

Yuri worked her way down the length of the cone as she reviewed this. When the cone had vanished, she gave a firm nod.

"This '$0.999\cdots$' has got to go. It's confusing," she said.

"How so?"

"Well think about it. Say you're writing a sequence of numbers like this."

$$0.9, \ 0.99, \ 0.999, \ \ldots$$

"You add those three dots at the end to mean 'goes on forever,' yeah? That made me think that when we were playing with $0.9, 0.99, 0.999$, a $0.999\cdots$ would eventually show up. But it doesn't, does it? There's no $0.999\cdots$ at the end of $0.9, 0.99, 0.999$ somewhere. We need a new way of writing this. Something like..."

- $0.9, 0.99, 0.999, \ldots$ becomes arbitrarily close to $\heartsuit$.

- $\heartsuit$ is therefore equal to 1.

"This would make things so much clearer."

I nodded. "I agree."

"But it's *your. Fault. Too*," Yuri said, punctuating her words with finger jabs at my chest. "You should have told me from the beginning that $0.999\cdots$ was a notational whatever, not a normal number. This problem isn't really about math, it's just about how you write things!"

"Yeah, I guess you're right. Glad to see you get it now, though."

"Well at least I didn't see this at school first. There's no way I'd have figured out what was going on. That $0.999\cdots$ isn't some number that's eventually going to show up in a sequence, it's the destination where the sequence is heading, even if it never gets there.

And *that's* why $0.999\cdots$ and 1 are absolutely, positively, right-on-the-nose the same."

I nodded again, smiling.

"Oh," Yuri said, "I just realized something else. These two numbers are different too, right?"

$0.999\cdots$	(equal to 1)
$0.999\cdots 9$	(less than 1)

"They are. The three dots at the end of a number show where the number is heading. The three dots in the middle of a number is just an abbreviation. The first one never shows up in the sequence, the second one will, eventually. They're completely different things."

"They're way confusing things, is what they are."

"I know you'll never confuse them again, though."

I looked down, and saw a white plastic bag at my feet. It took me a moment to realize what I was looking at.

"The eggs! Mom's still waiting for us!"

Answer 4-1

The following is a true statement.

$$0.999\cdots = 1$$

4.3   IN THE MUSIC ROOM

4.3.1   Introducing Variables

"You think it was Yuri's boyfriend that gave her that problem?" Tetra whispered.

We were hanging out in the music room after classes. Seated next to Miruka at the piano was her friend Ay-Ay, president of the school piano club "Fortissimo." Ay-Ay was in eleventh grade, like Miruka and me, but in a different homeroom. She was into music like I was into math, and spent most of her free time here, practicing. The music room was normally locked up after hours, but she was so talented that the music department head had given her a key.

We watched Miruka and Ay-Ay's backs as they played, Miruka's long, straight hair in sharp juxtaposition to Ay-Ay's wavy locks. They took turns playing pieces and argued between each. Ay-Ay would insist that they differentiate between 'mechanical Bach' and 'celestial Bach,' while Miruka argued the need to extract the 'formal Bach' from the 'meta-Bach.'

I had not a single clue what they were talking about.

"No way," I whispered back to Tetra, knowing that Miruka was in no mood to have her music interrupted.

"Betcha he is," Tetra said, a sly smile on her face. "Or wants to be. He's trying to impress her. It's a sign of affection."

I brushed away her words with a gesture.

"Yuri sure is smart," Tetra continued, a hint of admiration in her voice. "I still can't shake the feeling that $0.999\cdots$ is a little bit less than 1."

Tetra pulled her notebook and a pencil out of her bag.

"You said that when you explained all that to Yuri, you used n nines in the decimal. Like this, right?"

$$a_n = 0.\underbrace{999\cdots9}_{n\ 9s}$$

"Sure. Using variables like that can make explanations a lot clearer. In this case two variables, I guess—the a and its subscript n. Anyway, doing that's a lot easier than saying 'the number of nines in the number $0.999\cdots9$.'"

"I can see that. I need to get a lot more comfortable with introducing variables like this. It doesn't come natural to me at all. Seeing more letters still just makes everything look more confusing."

Tetra began writing letters in her notebook, as if practicing.

I looked back to the piano, where it was Ay-Ay's turn to play. Miruka had gotten up and was standing behind Ay-Ay with her arms crossed. She glanced back at me but immediately returned her gaze to the keys.

4.3.2 Limits

I motioned Tetra back to a further corner of the music room, where we could speak in more normal voices.

"Let's talk about limits, then," I said. I held out my hand, and Tetra passed me the notebook and pencil.

"Say you have some number a_n, and the bigger you make n, the closer a_n gets to some certain number. In a situation like that, the number you're getting close to is called a limit. You write it like this."

$$\lim_{n \to \infty} a_n$$

"Let's use A to name the number that a_n is getting closer and closer to. Then you can say that this limit equals A, which you write like this."

$$\lim_{n \to \infty} a_n = A$$

"You can also write it without using the limit notation, like this."

$$a_n \to A \qquad \text{as} \qquad n \to \infty$$

"A couple more words to learn. When a sequence can get arbitrarily close to some number, you say it *converges* to that number. So saying that a sequence converges is the same as saying it has a limit. Also, finding the limit of some sequence is often called *taking the limit*."

The limit of a sequence

a_n gets arbitrarily close to A as n increases

$\Longleftrightarrow \quad \lim_{n \to \infty} a_n = A$

$\Longleftrightarrow \quad a_n \to A \quad \text{as} \quad n \to \infty$

$\Longleftrightarrow \quad$ sequence $\langle a_n \rangle$ converges to A

I watched Tetra's eyes as she read what I'd written. She pointed at a line.

"How do you read this?"

$$a_n \to A \qquad \text{as} \qquad n \to \infty$$

"You'd say 'a_n approaches A as n approaches infinity.'"

"And this one?"

$$\lim_{n \to \infty} a_n = A$$

"I'd read that, 'the limit of a_n as n approaches infinity is A.'"

"Hmm... I think the first one is easier to understand, but the second one is the right one, right?"

"It isn't more correct, but it's what you'll see most often. It's more compact, if nothing else."

Tetra reflected on this for a moment, her eyebrows drawing in.

"Since we're talking about things getting closer, why the equals sign? Shouldn't it be this?"

$$\lim_{n \to \infty} a_n \to A \quad ?$$

"Ah, interesting," I said. "But no, the arrow shows change. We want to use it in the $n \to \infty$ part, to show that n is getting bigger and bigger. But if we wrote $\lim_{n \to \infty} a_n \to A$, that would mean the limit of a_n is getting closer to A."

"Isn't it?"

"No. Watch out for that. The limit $\lim_{n \to \infty} a_n$ is a specific number, the number that $\langle a_n \rangle$ converges to. That doesn't change."

$$\lim_{n \to \infty} a_n \quad \to \quad A \qquad \text{incorrect}$$
$$\lim_{n \to \infty} a_n \quad = \quad A \qquad \text{correct}$$

"Oops, sorry, I've got it straight now." Tetra said, blushing. She looked back at the notebook. "Oh, one more thing. Not all sequences will converge to something, right?"

"Absolutely not. What does a sequence like this do?"

$$10, 100, 1000, 10000, \ldots$$

"It keeps getting bigger and bigger and bigger." Tetra spread her arms wider and wider to illustrate.

"That's right, and it never stops getting bigger. There's no way it will get closer and closer to some specific number. We say a sequence like this doesn't *con*verge, it *di*verges. This particular sequence never stops getting bigger, so we can say that it diverges to positive infinity."

"We can't say that it *converges to* positive infinity?"

"No. Positive infinity isn't a number, so you can't get arbitrarily close to it. So you can never say that a sequence has infinity as a limit, or that it converges to infinity. You can just say that it diverges to positive infinity—or negative infinity, if it's heading that way."

"Okay, got it."

4.3.3 Sound Makes the Music

I heard voices coming toward us.

"I'm telling you, the C♯ just doesn't work there," Ay-Ay was saying.

"Oh, I don't know..." Miruka said.

"It breaks the pattern!"

"Maybe that's why my fingers don't seem to want to hit it."

The two walked up to where we had taken refuge. Ay-Ay's face was dark.

"Taking a break?" I asked.

"What are you two talking about?" Miruka said, ignoring me.

"Limits!" Tetra said. "This is some tricky stuff."

"Oh?" Miruka said, cocking her head.

"Well, I think I get the idea of getting arbitrarily close to something, but when you put it all in symbols I'm not so sure any more. I lose my instinct for it, somehow."

Ay-Ay stepped forward and inserted herself into the conversation.

"Listen," she said, "I'm no mathematician, but the way I see it, you use all those symbols in math because that's the best way to say what needs to be said."

Ay-Ay held out her hands and examined her palms. She turned them over and considered their backs for a time. I regarded her impressively long, strong-looking fingers—the fingers of a true pianist.

"Music is all about sounds," she continued, still looking at her hands. "Just...sounds." Her tone was uncharacteristically serious.

"Sometimes you can describe the world using words. When you can do that, great. Have at it. But there's also a world that can only be described with sound."

She formed a hand into a fist, thumb extended, and used that to point to her chest.

"Music is *mine*. It's what I use to let loose feelings that would rip me to shreds if I kept them pent up. It's all I *can* use. So I eat for music, I breathe for music, I live for music."

Ay-Ay's solemn air and grim expression left no room for response.

"Sometimes I meet people who say they 'don't get' music," she said. "They're usually the kind of person who can't 'get' anything they can't put into words, because music demands understanding on its own terms. Music isn't about words, it's about sounds. It can only *be* sounds. If you think you can express it in words, you aren't really listening. If you're searching for words, you aren't experiencing the performance, you aren't hearing the music. You're in a different time and space, where the music isn't. I want to scream at people like that, 'Stop looking for words! Open your ears!'"

Ay-Ay paused, took a deep breath, and looked at Tetra.

"If you try to study math without reading the equations, without *really* reading them, aren't you doing the same thing?"

Tetra let out a small gasp.

"Wow, I totally see what you mean," Tetra said. "If you don't read equations, you aren't seeing the world that mathematics is presenting. I guess if you try to stick to words without embracing the math, you aren't really *doing* math, you're just kind of watching it."

"Music and mathematics seem like completely different things on the surface," I said, "but in a lot of ways they're really similar."

Ay-Ay nodded.

"Musicians describe a world of music, so shut up and listen to the sounds. Mathematicians describe a world of mathematics, so shut up and embrace the equations."

Tetra smiled. "So sounds are the words of music, and equations are the words of mathematics!"

"It's always back to words with this girl," Ay-Ay grumbled.

"Oh, not literally!" Tetra rushed to add. "I just mean ... as a basic unit of representation."

"It's not only equations, though," I said. "It's concepts, too. Like, when we talk about limits, we say a value gets 'arbitrarily close to' some other number, not that it reaches it. I think a deep understanding of concepts like that are vital to understanding the equations that represent them."

"In any case," Ay-Ay said, "I'll stick to my music. I don't know if I'll be able to get a job related to music after I graduate, but it's something that will be part of me for the rest of my life. Absolutely."

Ay-Ay slapped her hands together, shattering the somber mood she'd created.

"Why're you guys all acting so *serious*?" she said, laughing. "Time to lighten up."

I smiled at her.

"I don't think you have anything to worry about," I said, "about working with music and all. Your playing is amazing, and your compositions are genius. I can't wait to see what you go on to do."

Ay-Ay pounded me on the back.

"You're a good kid," she said. "For a math nerd."

4.3.4 Calculating Limits

Ay-Ay left the room, saying she needed a break. The purely musical element of our set now gone, our talk soon turned to math.

"So did you show Tetra how to take some basic limits?" Miruka asked me.

"Like what?" I asked.

"Like this," she said, helping herself to my notebook and pencil.

Problem 4-2 (Basic limits)
$$\lim_{n \to \infty} \frac{1}{10^n}$$

"Uh, well ... no?" Tetra said, turning panicky eyes my way.

"Give me a shot at this?" I offered.

"Be my guest," Miruka said, smiling as she handed back the pencil and paper.

"Okay, Tetra. So what we want to find is the value of this expression."

$$\lim_{n \to \infty} \frac{1}{10^n}$$

Tetra nodded. "So we're looking for the limit of $\frac{1}{10^n}$ as n heads off toward infinity, right?"

"Right. In terms like you suggested before, we're looking for the value of the 'club' in this expression."

$$\frac{1}{10^n} \to \clubsuit \quad \text{as} \quad n \to \infty$$

"Got it! How do we start?"

"As always, with an explicit representation of this sequence. Like I always say, examples are the key to understanding. Anyway, the sequence looks like this."

$$\frac{1}{10^1}, \; \frac{1}{10^2}, \; \frac{1}{10^3}, \; \frac{1}{10^4}, \; \frac{1}{10^5}, \; \cdots, \; \frac{1}{10^n}, \; \cdots$$

"What we want to know," I continued, "is if $\frac{1}{10^n}$ is heading for some specific value as n gets bigger and bigger. And if it is, we want to know what that number is."

"Hmm..." Tetra said, her eyebrows knitting in concentration.

"Let me give you a hint—for now, just pay attention to the denominators."

$$10^1, \; 10^2, \; 10^3, \; 10^4, \; 10^5, \; \ldots, \; 10^n, \; \ldots$$

"Oh," Tetra said. "So they're increasing like this?"

$$10, \; 100, \; 1000, \; 10000, \; 100000, \; \ldots, \; 10^n, \; \ldots$$

"That's right. So as n gets bigger, what happens to 10^n?"

"It gets bigger, too. Like, *way* bigger."

"Exactly. That means we can say this."

$$10^n \to \infty \quad \text{as} \quad n \to \infty$$

I tapped what I'd written with my pencil.

"This means that as n gets bigger, the denominator of $\frac{1}{10^n}$ gets bigger, without limit. So what will happen to the fraction $\frac{1}{10^n}$ as its denominator keeps increasing?"

"The fraction... should get smaller and smaller, right?"

"Exactly. In fact, we can make it arbitrarily close to 0, just by making n big enough. That means we can say this."

$$\frac{1}{10^n} \to 0 \quad \text{as} \quad n \to \infty$$

"Putting this into the standard form for limits, we get this."

$$\lim_{n \to \infty} \frac{1}{10^n} = 0$$

"So we've shown that the limit exists, and its value is 0."

Answer 4-2 (Basic limits)

$10^n \to \infty$ as $n \to \infty$, so $\frac{1}{10^n} \to 0$. We therefore have that

$$\lim_{n \to \infty} \frac{1}{10^n} = 0.$$

"Huh..." Tetra said, chewing on a nail as she pondered this. "One question?" she said.

"You bet."

"This came up in how you explained the problem, right?"

$$10^n \to \infty \quad \text{as} \quad n \to \infty$$

"It did."

"And does that mean we can write this?"

$$\lim_{n \to \infty} 10^n = \infty$$

"Yep. Something wrong with that?"

"Well, it's just that—maybe I'm not understanding this right, but doesn't this mean you're saying that the limit of 10^n here is infinity?"

"Sure."

"But didn't you also say that we can't say the limit of a sequence is infinity? Just that it diverges to infinity?"

"Ah, right. Sorry, I didn't explain that well enough. You're right to be suspicious here, since infinity isn't a number. But what's going on is that we've expanded the definition of the $=$ operator to mean this."

$$\lim_{n \to \infty} 10^n = \infty \quad \Longleftrightarrow \quad 10^n \to \infty \quad \text{as} \quad n \to \infty$$

"Oh, okay" Tetra said. "So $\lim_{n \to \infty} 10^n = \infty$ is another way of saying that the sequence $\langle 10^n \rangle$ diverges to infinity."

I nodded. "Yes, that's right."

An impatient Miruka broke her silence.

"Next problem."

Problem 4-3 (basic limits)

$$\lim_{n \to \infty} \sum_{k=1}^{n} \frac{1}{10^k}$$

Tetra looked back and forth between this problem and the previous one.

"Um . . . is this not the same thing as what we just did?" she asked.

Miruka grimaced. "Who was it that was just going on about how not reading equations prevents you from seeing the world that mathematics is presenting?"

"Okay, let me read this one more time, carefully." Tetra pulled the notebook closer and peered at the problem. "Oh, I get it," she said. "The sigma makes all the difference, doesn't it. I have no idea how to solve this, though. How do you take the limit of a sum?"

"Have fun," Miruka said, patting me on the shoulder. She turned and headed back to the piano.

"Okay," I said. "Here's what we want to solve."

$$\lim_{n \to \infty} \sum_{k=1}^{n} \frac{1}{10^k}$$

"To do that, first we need to pay attention to exactly what it is we're taking the limit of."

$$\sum_{k=1}^{n} \frac{1}{10^k}$$

"Let's think about how we can represent this as an expression involving just n. That'll be a lot easier than dealing with the sigma. So what's the first thing to do to make sure you understand what's going on here?"

"I know! Write out some examples!"

Tetra took the pencil and started writing.

$$\sum_{k=1}^{1} \frac{1}{10^k} = \frac{1}{10^1} \qquad \text{(for } n = 1\text{)}$$

$$\sum_{k=1}^{2} \frac{1}{10^k} = \frac{1}{10^1} + \frac{1}{10^2} \qquad \text{(for } n = 2\text{)}$$

$$\sum_{k=1}^{3} \frac{1}{10^k} = \frac{1}{10^1} + \frac{1}{10^2} + \frac{1}{10^3} \qquad \text{(for } n = 3\text{)}$$

"Very good," I said. "Can you use this to write a general expression?"

"I think so... Yeah, I can!"

$$\sum_{k=1}^{n} \frac{1}{10^k} = \frac{1}{10^1} + \frac{1}{10^2} + \frac{1}{10^3} + \cdots + \frac{1}{10^n} \qquad \text{(general expression)}$$

"Good job—now we've established the groundwork for finding this limit. Next we want to massage this into a more useful form, one with the terms shifted. We can do that by multiplying both sides of the equation by $\frac{1}{10}$."

$$\sum_{k=1}^{n} \frac{1}{10^k} = \frac{1}{10^1} + \frac{1}{10^2} + \frac{1}{10^3} + \cdots + \frac{1}{10^n} \qquad \text{general expression}$$

$$\frac{1}{10} \cdot \sum_{k=1}^{n} \frac{1}{10^k} = \frac{1}{10} \cdot \left(\frac{1}{10^1} + \frac{1}{10^2} + \frac{1}{10^3} + \cdots + \frac{1}{10^n} \right) \qquad \text{mult. both sides by } \frac{1}{10}$$

$$\frac{1}{10} \cdot \sum_{k=1}^{n} \frac{1}{10^k} = \frac{1}{10} \cdot \frac{1}{10^1} + \frac{1}{10} \cdot \frac{1}{10^2} + \frac{1}{10} \cdot \frac{1}{10^3} + \cdots + \frac{1}{10} \cdot \frac{1}{10^n} \qquad \text{expand the right side}$$

$$\frac{1}{10} \cdot \sum_{k=1}^{n} \frac{1}{10^k} = \frac{1}{10^2} + \frac{1}{10^3} + \frac{1}{10^4} + \cdots + \frac{1}{10^{n+1}} \qquad \text{expr. with terms shifted}$$

Tetra ran a finger down each line, confirming what I was doing in each step.

"By 'shifted terms,' you mean that the exponents on the 10s are each increased by 1, right?"

"That's right," I said. "Now we can subtract this shifted equation from the generalized equation. That will kill off all the intermediate terms."

$$\sum_{k=1}^{n} \frac{1}{10^k} = \frac{1}{10^1} + \frac{1}{10^2} + \frac{1}{10^3} + \cdots + \frac{1}{10^n} \qquad \text{general expr.}$$

$$-\quad \frac{1}{10} \cdot \sum_{k=1}^{n} \frac{1}{10^k} = \qquad \frac{1}{10^2} + \frac{1}{10^3} + \cdots + \frac{1}{10^n} + \frac{1}{10^{n+1}} \qquad \text{shifted expr.}$$

$$\left(1 - \frac{1}{10}\right) \cdot \sum_{k=1}^{n} \frac{1}{10^k} = \frac{1}{10^1} \qquad\qquad\qquad - \frac{1}{10^{n+1}} \qquad \text{difference}$$

"Oh, neat!" Tetra said. "Everything except the first and last terms went away!"

"Let's calculate this and see what we get."

$$\left(1 - \frac{1}{10}\right) \cdot \sum_{k=1}^{n} \frac{1}{10^k} = \frac{1}{10^1} - \frac{1}{10^{n+1}} \qquad \text{previous equation}$$

$$\frac{10-1}{10} \cdot \sum_{k=1}^{n} \frac{1}{10^k} = \frac{1}{10^1} - \frac{1}{10^{n+1}} \qquad \text{calculate left side}$$

$$\frac{9}{10} \cdot \sum_{k=1}^{n} \frac{1}{10^k} = \frac{1}{10^1} - \frac{1}{10^{n+1}} \qquad \text{simplify left side}$$

$$\sum_{k=1}^{n} \frac{1}{10^k} = \left(\frac{1}{10^1} - \frac{1}{10^{n+1}}\right) \cdot \frac{10}{9} \qquad \text{multiply both sides by } \frac{10}{9}$$

$$= \frac{1}{10^1} \cdot \frac{10}{9} - \frac{1}{10^{n+1}} \cdot \frac{10}{9} \qquad \text{distribute}$$

$$= \frac{1}{9} - \frac{1}{9 \cdot 10^n} \qquad \text{simplify}$$

I rechecked my work. Satisfied, I nodded.

"Now we need to think about what's going to happen to the right side here as n goes to infinity."

$$\sum_{k=1}^{n} \frac{1}{10^k} = \frac{1}{9} - \frac{1}{9 \cdot 10^n}$$

"Hmm," Tetra said, putting a finger to her lips. "When n goes to infinity, the limit of the $\frac{1}{9 \cdot 10^n}$ part here will be 0, won't it?"

"It will. That means we can say this."

$$\sum_{k=1}^{n} \frac{1}{10^k} \to \frac{1}{9} \quad \text{as} \quad n \to \infty$$

"In other words . . . "

$$\lim_{n \to \infty} \sum_{k=1}^{n} \frac{1}{10^k} = \frac{1}{9}$$

Answer 4-3 (basic limits)

$$\lim_{n \to \infty} \sum_{k=1}^{n} \frac{1}{10^k} = \frac{1}{9}$$

"You guys done?"

I glanced back and saw Miruka standing behind us, holding some sheet music. I nodded.

"Calculate $0.999\cdots$ next, then."

Problem 4-4

Calculate $0.999\cdots$, defining $0.999\cdots$ as follows:

$$0.999\cdots = \lim_{n \to \infty} 0.\underbrace{999\cdots9}_{n \ 9s}$$

I blinked at this unexpected problem, but soon laughed.

"So *this* is what you've been leading us to."

"You finally noticed," she said, taking the notebook and solving the problem herself.

$$0.999\cdots = \lim_{n\to\infty} 0.\underbrace{999\cdots 9}_{n\ 9\text{'s}}$$

$$= \lim_{n\to\infty} \left(0.9 + 0.09 + 0.009 + \cdots + 0.\underbrace{000\cdots 0}_{(n-1)\ 0\text{'s}} 9 \right)$$

$$= \lim_{n\to\infty} \left(\frac{9}{10^1} + \frac{9}{10^2} + \frac{9}{10^3} + \cdots + \frac{9}{10^n} \right)$$

$$= \lim_{n\to\infty} 9 \cdot \left(\frac{1}{10^1} + \frac{1}{10^2} + \frac{1}{10^3} + \cdots + \frac{1}{10^n} \right)$$

$$= \lim_{n\to\infty} 9 \cdot \sum_{k=1}^{n} \frac{1}{10^k}$$

$$= 9 \cdot \lim_{n\to\infty} \sum_{k=1}^{n} \frac{1}{10^k}$$

$$= 9 \cdot \frac{1}{9} \qquad \text{from Answer 4-3}$$

"And thus, $0.999\cdots$ equals 1," Miruka said.

Answer 4-4

$$0.999\cdots = \lim_{n\to\infty} 0.\underbrace{999\cdots 9}_{n\ 9\text{s}} = 1$$

"Wow," Tetra said. "So you can just, like, calculate $0.999\cdots$ out?"

"Yeah," I said, "if you're clever about how you define things."

"Infinity fools the senses," Miruka said. "If you try to rely on common sense when you deal with infinity, you'll get tripped up every time. Not all of us can be an Euler."

"I see," Tetra said.

"So don't rely on your senses, rely on—" Miruka looked at me.

"Logic," I said.

Miruka turned to Tetra.

"Don't rely on words, rely on—"

"Equations."

Miruka smiled, and Tetra raised her hand.

"And that's why we use this lim operator, instead of words like 'gets closer and closer to' and all, right?"

Miruka gave a reluctant nod.

"Yes, but the way we've been treating limits so far, the lim operator isn't much better than just a word."

"Why not?"

"Because we haven't defined it. Not mathematically, at least." Miruka began walking slowly around us. "At some point we're going to have to leave the words behind."

"But . . . but how?" Tetra asked.

"By using equations instead, of course," Miruka replied.

"You can use equations to *define* limits? Not just find them?"

"Now we can," Miruka replied with a grin. "But actually that's a relatively new development. Cauchy first brought rigorous concepts of limits into mathematics in the early 1800s, but it wasn't until late in that century that Weierstrass finally gave us a full definition using equations."

The door creaked as Ay-Ay reentered the room.

"Break's over!" she announced. "Miruka! Back at it!"

Tetra was still mumbling to herself, a perplexed look on her face.

"Limits . . . ? With equations . . . ?"

Miruka playfully bopped Tetra's head as she headed back toward the piano.

"We'll get there soon," she said, smiling. "To the realm of epsilon–delta."

4.4 HEADING HOME

4.4.1 Moving on

Knowing that Miruka would still be practicing for some time, Tetra and I headed off for the station. Tetra walked a half-stride behind me. The chilly air carried a faint scent of plum blossoms.

"Intense talk today, huh," Tetra said.

"Yeah," I agreed. I recalled our conversation in the music room, and Ay-Ay's deadly seriousness when it came to music.

"So what do you think you want to do in the future?" Tetra asked.

"I dunno. How about you?"

"I've always thought I wanted to find a job related to languages, but recently I've been really enjoying studying computers too." Tetra shrugged. "We'll see where my studies take me, I guess. I kind of envy Ay-Ay for knowing what she wants with so much certainty."

"Me too."

Tetra will end up an interpreter or a translator—she's too good at languages to give them up. I wonder how serious Yuri is about being a lawyer. Not very, most likely. Then again, maybe that would suit her—

"—don't you think?" Tetra was saying.

"Hmm? Uh, yeah, sure."

And what about Miruka? I can't imagine her as anything but a mathematician, but who knows? She's smart enough to do pretty much anything she wants...

Some change in the air brought me back to earth, and I realized I was walking alone. I turned to see Tetra, standing and looking down at the street.

"What's wrong?" I asked, but Tetra remained still and silent.

I walked back to her. I leaned over to get a better look at her face.

"You okay?" I prodded.

When she answered I could hardly hear her.

"I...I've got nothing," she said.

"Huh? What do you mean?"

"Miruka's a whiz at math, Ay-Ay can make such beautiful music, but I—what do I have?"

Tetra dragged a sleeve across her eyes before continuing.

"I've learned to like math, thanks to you, but it's not like I can really *do* it. All I do is waste your time with my silly questions, and never give anything in return. I'm sor—"

"No, Tetra, no. You've given me a lot. More than you'll ever know. You've taught me how to be more patient, for one thing. Now whenever I get frustrated with a problem and I just want to quit, I think of you—how hard you work—and I find the strength to keep at it."

Tetra eyes remained fixed on the ground.

"So enough of that stuff, okay? I want you to keep asking me questions, anything and any time. I learn better that way."

Tetra finally looked up.

"Okay," she said, nodding grimly. "Yeah, okay. I will. Thank you."

I nodded in return, and we started back down the road.

"But seriously, let me know any time I'm being a bother," she said, placing a hand on my arm. "You have more important things to worry about."

The destination toward which the sequence $1, \frac{1}{2}, \frac{1}{3}, \ldots, \frac{1}{n}, \ldots$ is heading is called its *limit*, which we write as $\lim_{n \to \infty} \frac{1}{n}$. We furthermore say that this sequence *converges to* 0. But beware! We are only using the word 'limit' to indicate where the sequence is heading. We never, ever mean that the sequence will reach that goal after you're done 'performing an infinite number of operations.'

NORIO ADACHI
The Paradox of Infinity

CHAPTER 5

Leibniz's Dream

Validity need have no relation to time, to duration,
to continuity. It is on another plane, judged by other
standards.

ANNE MORROW LINDBERGH
Gift from the Sea

5.1 YURI IMPLIES NOT TETRA

5.1.1 *The Meaning of 'Implies'*

"I don't get 'implies,'" Yuri said as she entered my room the next Saturday.

I looked up from my desk.

"And a good morning to you, too," I said.

"No time for pleasantries. What's this 'A implies B' thing all about? You know, in logic."

I sighed. "If you really want to understand it, we have to start at the beginning."

"Before we start, you *do* know what it means, right?"

"Sure."

"Just checking," Yuri said with a wink.

I sighed again, shaking my head, and opened a notebook to a blank page. Yuri pulled a chair closer to my desk and plopped down into it. She put on her glasses, a sure sign she was getting down to business.

"Okay," I began. "Say you have two propositions, A and B. We can use the 'implies' operator—an arrow—to connect them and create a new proposition, like this."

$$A \Rightarrow B$$

"Yep yep," Yuri said, nodding.

"Remember that a proposition is a mathematical statement that is either true or false. So if you have two propositions, there are four true–false patterns they can be in, and we can evaluate the truth of the proposition $A \Rightarrow B$ for each case. We can summarize all that using something called a truth table."

A	B	$A \Rightarrow B$
F	F	T
F	T	T
T	F	F
T	T	T

"Yes!" Yuri said. "Yes! This is exactly what I don't get!"

"You don't get how to read a truth table?"

"Of course I can read this. The first row says that if A is false and B is false, then $A \Rightarrow B$ is true, right?"

"Yeah, you've got it. So what do you not understand?"

"The first and second lines. I'm good with the other two."

"I'm not sure what—"

"Just look at it," Yuri said. "Think about what 'implies' means. A is false in both of these rows, yeah? So how can you even have an implies relationship in that case? It's broken from the get-go, but the truth table says that A does imply B. *That's* what I don't get."

"Okay, hang on, let me think about how to explain this..."

"Well it better be a good explanation, because I told this guy there's no way it makes sense, no matter how much he insisted it did. Not like he could explain it, either..."

'This guy'?

"How about this—what kind of truth table would make sense to you?"

Yuri narrowed her eyes, suspicious.

"What do you mean?"

"Well you said the third and fourth lines make sense, so I guess we can leave those alone. That means we can just check out all the true–false possibilities for the first two rows, by setting them up in a new table. Then you can tell me which pattern you think makes the most sense."

"See? This is why you're my go-to guy for stuff like this."

"Hang on, lemme make a table . . . "

A	B	(1)	(2)	(3)	(4)
F	F	F	F	T	T
F	T	F	T	F	T
T	F	F	F	F	F
T	T	T	T	T	T

"That's all the possibilities?" Yuri asked, leaning forward on my desk.

"Back off, I can't see," I said. "And yes, that's all. So which of these columns (1) through (4) do you want to show the 'implies' relationship?"

"One where saying that A is false means that you can't say anything about what it implies."

"In other words, if A is false you want 'A implies B' to be false, too. In that case, you have to use column (1)."

"Right! Much better!"

"Well hang on. If that's the case we only get 'true' when both A and B are true. So column (1) here is what's called the AND operator—A AND B."

"Ah, okay. Yeah, I guess it wouldn't do to have two different operators that give the same answer. Hmm . . . "

"Also, (2) is just B itself. It just takes the value of B, regardless of what A is, so there's no point in having an operator for that."

"Hey, you're right. So what's (3), then?"

"That one's $A = B$. It's true when A and B are both either true or false."

"Okay, yeah, I see that. We don't want 'implies' to mean 'equals,' I guess..."

"Which leaves us with just (4), since none of the other columns seem usable. My advice is to not get too hung up on the word 'implies,' and just think of it as an arbitrary name."

"Consider me reluctantly convinced. There's no arguing with a truth table, am I right?"

5.1.2 Solving without Thinking

"You sure have a knack for logic," I said. "I don't think there's many kids your age that would get this stuff as well as you do."

"I have my moments," Yuri said.

Yuri went to my bookcase and started rummaging through my library. She pulled a book off a shelf that I remembered being out of her reach not so long ago.

"Look for a book about Leibniz," I suggested. "You'd like him. He's the one that turned logic into equations."

"Was that a recent thing?"

"Not really. He was a contemporary of Newton. Seventeenth century, I guess that would make it. You know Newton, right?"

"Sure. He was that gardener that studied how apples fall."

"No! He was the physicist who discovered the law of universal gravitation!"

Yuri rolled her eyes and shook her head.

"Oh, you knew that," I said. "Never mind, just making sure." I cleared my throat, and continued. "Anyway, Leibniz believed thought could be reduced to series of calculations. He wanted to find a way to mechanically calculate out logical thought."

"You mean he wanted to build a thinking machine? Like a computer?"

"In modern terms, yeah, something like that. He once predicted that the day would come when we would learn how to use logic to, like, calculate all truths. That nobody would ever get into arguments, because it would be possible to just use equations to figure out who was right."

"So *that's* the solution to world peace! Everyone just needs to study more math, so we can all argue in equations!"

"Yeah, he was something of an optimist. But it's still an interesting idea—solving problems without thinking about what they mean."

"Hey, check this out," Yuri said. She contorted her face into an expression somewhere between bewilderment and horror.

"What's that supposed to be?" I asked.

"How I'm going to react on the day you suggest that I solve a math problem without thinking about what it means."

I tried—and failed—to resist laughing.

"What I'm talking about," I said, "is how you should approach manipulating equations. Sometimes, thinking too much about what you're doing can make it easy to mess up. Like, when you started studying basic algebra, didn't your teacher keeping harping about how you need to set things up as equations?"

"Man, don't you know it. Even stupid easy stuff that didn't need it. Half the time the answer would just pop into my head, so I'd write that down first, then come up with an equation that fit."

"What you were *supposed* to be learning is how to do your thinking up front when you're setting up mathematical statements that represent the problem. Then you can just kick back and solve for x or whatever, without really worrying about what every step means. Sure, it's important to understand what the problem says, but once the math is set up you've taken things from the world of meaning to the world of equations. Once you're there you can just use the tools and methods of math to manipulate symbols without worrying about what each one stands for. Take the results back into the real world, and there's your answer, with meaning again."

"I have *no* idea what you're talking about."

"Sure you do. Say you're doing a problem where you have to find out how much some apples cost. When you've said the apples cost x cents each, you've taken things into the world of equations. After you've done some algebraic shuffling about and find that they cost 40 cents, you've come back into the world of meaning. The world of equations is like a mirror, reflecting the real world. When you hold the mirror just right, it can show you how to solve real-world problems using equations."

I drew a quick sketch to illustrate what I was trying to say.

The world of meaning   The world of equations

Price of apple $\xrightarrow{\text{price is x cents}}$ Equation in x

$\downarrow$ solve equation

40 cents $\xleftarrow{\text{x represents price}}$ x $= 40$

Solving a problem by passing it through the world of equations

Yuri crossed her arms, clearly dissatisfied.

"I don't think the world is quite that simple," she said.

"Well, admittedly the trick is in the setup, which won't always be as easy as pricing fruit."

"So what it comes down to is that if you can set your problems up as math, you can use math to solve your problems." Yuri shook her head. "Nice lecture, Dr. Obvious."

5.1.3   The Limits of Reason?

Yuri stretched her arms over her head and gave a half yawn, half yowl.

"Actually, that reminds me of something," she said. "You ever hear of something called Gödel's incompleteness theorems?"

"Sounds familiar, but I can't claim to know the details," I said.

"Huh. There's this guy who's really into math that I've been talking to—the same one that was telling me about 'implies' and all. Anyway, the other day he started talking about this Gödel's incompleteness theorem thing..."

"What about it?"

"That it was some advanced theory or something that proves mathematics is incomplete, I think? That math is the most precise area of science humans have managed to come up with, but that this theory shows how even that's flawed, that there are limits to human reasoning, blah blah blah. He got all worked up about it when we were talking after school. I didn't understand half of what he was going on about."

"You two meet and talk after school?"

"Sometimes. So anyway, when I told this guy to put up or shut up, to explain in detail what all that was about, he said he didn't really understand all the details. I was hoping you could fill me in, so I could show him up."

"How long were you talking after school?"

"I dunno. Long enough for him to fill the chalkboard up with all kinds of crazy graphs. He isn't as good as you at explaining math stuff, but it was kinda fun."

Oh to be a fly on that wall.

5.2   TETRA IMPLIES NOT YURI

5.2.1   *Test Anxiety*

On the way to school the following Monday, I was startled by a loud "Good morning!" I turned to see Tetra, waving and heading toward me.

"Hey. How's it going?" I said.

Tetra hesitated before answering.

"Okay, I guess."

"That's not like you. Something wrong?"

"Nothing's *wrong*, really. Just a premature case of test anxiety, I suppose."

"Ah, that."

It was February, which meant the seniors were buckling down, studying for their college entrance exams. Worrying about getting into a good university was still a year away for me—two years for Tetra—but the sudden seriousness of the seniors put the whole school on edge. We just tried to lay low and stay out of their way, to somehow make it through to spring.

"Something in particular got you worried?" I asked.

Tetra shrugged.

"They just seems so ... *big*. Normal school tests—even finals—are limited, you know? Like, I pretty much know what's going to be on them, so at least I know what to study for. But entrance exams ... *Anything* can be on those. High school entrance exams were bad enough. I just kept studying the same stuff over and over, filling up notebook after notebook. My friends all seemed to get stuff so much easier. I had to just hammer it all in."

I nodded without speaking. Tetra sighed deeply, and continued.

"But hey, it was worth it, right? I got into this school, so I could get you to teach me math, for one thing. I think I'm slowly developing the knack for studying it."

"There's a knack?"

"Sure. Thinking precisely, paying attention to definitions, reading things carefully ... "

"Ah, right. That kind of stuff."

"I enjoy studying English, so I guess I'll do okay on that part of the test. Maybe. Still, I worry about studying for exams. It feels like I'm supposed to do something special to prepare for them, but I'm not sure what."

We came to a red crossing signal, and stopped to wait.

I turned to Tetra and said, "Is all this really about college entrance exams? You're a couple years early to get so worked up about them, I'd think."

Tetra blinked several times, then started chewing on a fingernail as she thought.

"I suppose you're right," she said. "Really I'm just a mess when it comes to tests in general."

She fell into silence again, so I prodded.

"How, specifically?" I asked.

"Umm ... I just get flustered. I'm not good at managing my time. I give up on problems and move on to the next. I worry about running into a problem that I have no idea how to solve. That kind of thing."

"Maybe you just need more practice? Like, practice doing problems within a certain time, like you have to do when you take a test."

"Hey, that's a neat idea. I've never tried that."

"There's a time for thinking things through slowly, but sometimes speed is important too."

The signal turned green, and we began walking again.

5.2.2   Classes

We entered the warren of back streets that were a shortcut to our school.

"Next year you'll have entrance exam prep classes," I reminded her, "so that'll give you a chance to learn the basics of taking those tests. Of course, that's not enough. You'll need to understand what's being taught in your normal classes too."

"If only there was a trick for being sure I could do that," Tetra said.

"The trick is just to pay attention, to listen to what's being said. Not that I think you have a real problem doing that. But beyond that, you have to be sure you *understand* what's being said. Filling up notebooks and actually understanding what you're writing are two different things. You say you tend to get distracted by questions that pop into your head during class, right?"

"Oh, I totally do!"

"Maybe instead of worrying about things like that during class, you should just jot them down in your notebook as something to look into later, when you have more time. If you're paying attention in class instead of thinking about other stuff, that might be enough to answer your questions."

"Yep, you're absolutely right."

"But that aside, when it comes to preparing for entrance exams . . . I'm not sure what to tell you. I pay attention in class, read my textbook, review my notes . . . then back to step one. Of course, I try to always be thinking about what I'm doing, rather than just going through the motions."

"Yeah, you're always telling me how important it is to think things through."

"Sure. New material has to be chewed, not swallowed. If you don't take your time and think things through, you can't be sure you *really* understand it. Not that you have to wait for total comprehension before you move on to something else—the edges of understanding tend to be fuzzy. What's important is to never pretend you know something—you have to know where the gaps are. It's filling in those gaps that makes studying fun."

Tetra nodded silently.

"Don't use others as a basis for what you 'should' understand," I continued. "You've gotta be brave enough to see what you aren't getting, and to take the time to think things through until you've mastered it. Once you've done that, it's yours forever. And that's a pretty nice feeling. I think it would take you a long way toward not getting so worked up about tests."

Tetra nodded several more times.

"I'm sorry," I said. "I should shut up now."

"No, I need to be reminded of things like that from time to time," Tetra said. "And you're the only person I know who'll do it. My teachers sure don't. Neither do my parents. You're..."

Tetra looked down and away before continuing.

"You're...very important to me."

"Glad to be of help."

We passed through the school gates, and reached the point where we parted ways to head to our respective classes.

"See you after school?" I asked, turning to Tetra. She just stood there, shifting her weight.

"You okay?" I said.

Tetra turned her huge eyes directly at me.

"I—I—," she stuttered.

The warning bell rang, and Tetra seemed to somehow deflate.

"I'll see you after school," she said. She spun off and headed toward her building.

5.3   Miruka Implies Miruka

5.3.1   *In My Classroom*

When the last bell rang, I wandered back to Miruka's desk. She was leaning back, her attention focused on a book.

"Whatcha reading?" I asked, sliding into the desk across from her.

Still reading, Miruka lifted the book up so I could see the cover.

Gödel's Incompleteness Theorems

"Huh, what a coincidence," I said. "Yuri was talking about that just the other day. Something about it being a proof of the limits of reasoning."

Miruka peered at me over the book, a suspicious look on her face.

"*Yuri* was talking about this?" she asked.

"Yep."

Miruka's eyes narrowed in an imposing stare, causing me to shrink back into my seat.

"And you corrected her, right? About how that's not what this is at all?"

"I don't know enough about it to say anything, really."

I gave Miruka a rundown of what Yuri and I had discussed. Miruka didn't seem impressed.

"Leibniz's dream, huh? Hmph."

She set her book aside, closed her eyes, and sat silently. When she did this I would fall silent, too, knowing that something was brewing inside of her, something that I was eager to see come bubbling forth. Or maybe it was just that I loved the chance to stare at her face. So calm, so—

"Hey, guys! What's up?" came a familiar call from the door.

I tore my eyes from Miruka.

"Hey, Tetra."

"Oops, looks like Miruka's doing a deep think," Tetra said, covering her mouth with a hand. "I'll be quiet!" she stage whispered.

Despite the commotion, Miruka remained still, eyes closed, lost in thought. Tetra poked me, making a questioning face and pointing at the book that now lay face down in front of Miruka. On its back cover was a curious mark, apparently made with a rubber stamp.

As I was shrugging at Tetra, Miruka opened her eyes and spoke.

"It says 'Narabikura Library,'" she said. She turned to Tetra. "Glad you've joined us. It's time to play with formal systems of propositional logic."

5.3.2 Formal Systems

Miruka walked to the blackboard. Tetra and I followed and sat in the front row of desks.

She picked up a piece of chalk and began writing as she spoke.

"There are two main approaches to studying logic: using semantics, and using syntax. Semantics uses values like true and false to explore the relationship between propositions. But today I want to use syntax. With syntax we aren't so interested in true and false—instead we focus on the structure of logic statements. In other words, we're thinking about *forms*, not *meaning*."

Methods of studying logic

Semantics	Uses true/false values
Syntax	Does not use true/false values

"We use syntax to investigate formal systems. Let's use 'H' to name the formal system we're going to develop."

Formal system H

Tetra raised her hand.

"Um, Miruka? I have no idea what a formal system is. It sounds awfully...abstract. I don't even know what I'm supposed to be imagining!"

"No worries," Miruka said, smiling. "Just hold on until I get an example up here."

Tetra gave an apprehensive okay, and Miruka continued.

"We have to start out by defining some concepts. Here's the order we'll do that in."

· Logical expressions

· Axioms and inference rules

· Proofs and theorems

"Axioms, proofs, theorems... These are all words we know well from mathematics, but now we're going to define them for formal systems. That will give us the formal system H we're after, which we'll be able to manipulate as a miniature model of mathematics."

"A miniature—?" Tetra began, but Miruka cut her off with a wave of a chalky hand.

"This is the first step toward doing math on mathematics."

"Doing math on..." Tetra trailed off. From the look on her face, I assumed she was so baffled she couldn't even finish repeating what she'd heard.

I'd have been the same, if I'd had the nerve to speak myself. I had no idea where Miruka was going with all this.

"Enough rhetoric," Miruka said. "Let's talk formulas."

5.3.3 Propositional Formulas

Miruka began writing on the board again.

"We're going to define propositional formulas in our formal system H like this," she said.

Propositional formulas (Formal system H, definition 1)

Only rules F1–F3 define formulas:

Rule F1 If x is a variable, x is a formula.

Rule F2 If x is a formula, $\neg(x)$ is also a formula.

Rule F3 If x and y are formulas, $(x) \vee (y)$ is also a formula.

Miruka looked over what she'd written and nodded.

"For the variables mentioned in F1 here, we'll use capital letters like A, B, C. We don't want there to be a finite number of possible variable names, though, so we'll say that subscripts like A_1, A_2, A_3 can be used, too."

Miruka wrote something on the board.

> Is A a formula?

She pointed at Tetra, and said, "Show me you're keeping up. What's the answer?"

"Um, yes, I think so," Tetra said.

"Why?"

"*Why*? Because, uh ... I'm not sure what to say."

"Just say why A is a formula. It's because A is a variable, and F1 says that if x is a variable, then x is also a formula. Therefore A is a formula."

"Oh, so I just have to cite the definitions. Got it."

"Next problem," Miruka said.

> Is ¬(A) a formula?

"Yes, it is," Tetra promptly replied.

"Why?"

"Because A is a formula, and F2 says that if x is a formula, then ¬(x) is, too. If we let x be A, then we can create ¬(A)."

"Well done. Next."

> Is (A) ∧ (B) a formula?

"Yes," Tetra said.

"Wrong," Miruka shot back. "There's no logical AND symbol anywhere in the definitions. That's not AND in F3, it's OR. So (A) ∧ (B) is *not* a formula in our formal system H."

"Agh ... Pay attention! Pay attention!" Tetra said, pounding the heel of her hand against her forehead.

"New problem," Miruka said.

> Is A ∨ B a formula?

"Sure!" Tetra said. "Because this time we're definitely using the OR symbol!"

"Wrong," Miruka said. "Note the parentheses."

A ∨ B	An invalid formula in formal system H
(A) ∨ (B)	A valid formula in formal system H

Tetra's eyes moved between what Miruka had written on the board and the original definition.

"Oh, I see. So we aren't allowed to just skip the parentheses."

"There is a way, but let's save that for later. I want to emphasize the syntax that says the order of things is important, so for now we'll stick to explicit parentheses."

"Got it," Tetra said, jotting something down in her notebook.

"Next problem," Miruka said.

> Is $(\neg(A)) \lor (A)$ a formula?

"Wow, they're getting tricky. Umm ... yes, I think it is."

"Why?"

"Well, $\neg(A)$ and A are formulas. If we let x be $\neg(A)$ and let y be A, then we can apply F3, which says that if x and y are formulas, $(x) \lor (y)$ is, too."

"Well done. Next."

> Is $\neg(\neg(\neg(\neg(A))))$ a formula?

"Uh ... wow."

Tetra leaned forward and began carefully counting parenthesis.

"One, two, three, four ... Yes. Yes, this is a formula."

"Right. How'd you get that?"

"F2 says that if x is a formula, then $\neg(x)$ is, too. You just have to use that rule over and over."

Tetra went to the blackboard and picked up a piece of chalk.

A	1. This is a formula (from F1)
$\neg(A)$	2. This is a formula (from 1 and F2)
$\neg(\neg(A))$	3. This is a formula (from 2 and F2)
$\neg(\neg(\neg(A)))$	4. This is a formula (from 3 and F2)
$\neg(\neg(\neg(\neg(A))))$	5. This is a formula (from 4 and F2)

"This feels a lot like successors in Peano arithmetic," I said.

"It does, doesn't it!" Tetra said.

"Because it's a recursive definition," Miruka said. "We're using propositional formulas to define propositional formulas."

5.3.4   The Form of 'Implies'

"Let's make formulas in our formal system H a little easier to read," Miruka said. "We can do that by defining the 'implies' symbol."

The → symbol (Formal system H, definition 2)

'Implies' symbol   Define $(x) \to (y)$ as $(\neg(x)) \lor (y)$.

"This says we'll be using $(x) \to (y)$ as a shorthand for $(\neg(x)) \lor (y)$."

"Can you show me an example?" Tetra asked.

"Here's a simple one," Miruka replied.

$$(A) \to (B)$$

"Writing this is the same as writing this."

$$(\neg(A)) \lor (B)$$

"Okay," Tetra said, nodding.

"Can you rewrite this without using the implies symbol?"

$$(A) \to (A)$$

"I think so."

$$(\neg(A)) \lor (A)$$

"That's right."

"And $(A) \to (A)$ will always be true, right?" Tetra asked.

"What do you mean, true?" Miruka said, a gleam in her eye.

"What do I mean? Er...just that. That 'A implies A' is always a true statement."

"But we're talking about a formal system. There *is* no true or false."

"Oh!" Tetra said. "That which isn't allowed is forbidden!"

"What do you mean?" Miruka asked.

"That later on we might define things like 'true' and 'false,' but we can't assume meanings like that until we've done so. That even if we know where we're headed, we can't use the tools we'll find there until we've arrived."

Miruka grunted in agreement. "That sums it up, I suppose. But more to the point, when you're dealing with a formal system, you want to be cool and mechanical. Don't try to extract meaning. When you see a formula like $(\neg(A)) \vee (A)$, don't think of it as trying to say something. Just accept it as a string of separate symbols, like keys on a keyboard."

"Why do we need to do that?" Tetra asked.

"Because thinking clouds reason," Miruka said. "A proof performed in the absence of thought, one that utilizes only the form of previous definitions, has a better basis of rationale."

Now I know where all those whys are coming from. She's checking Tetra's basis for her answers.

Tetra responded with silence, clearly considering Miruka's words. She really seemed to be taking this all to heart.

"But, still..." she said. "All we've done is say that $(A) \to (A)$ is an abbreviation of $(\neg(A)) \vee (A)$. That seems too simple to be called a proof."

"Because we've only defined what a formula is," Miruka said. "Let's move on to axioms."

I felt something big stirring within me, listening to them talk. Miruka had said we're aiming at creating *a miniature model of mathematics*—something that seemed so profound as to be beyond our reach. Before, when we talked about Peano arithmetic, we'd defined things like the natural numbers and how to add them. At the time it had felt like we were looking at the very foundation of mathematics. But this formal system H—this was a whole level deeper, the foundations of the foundations. A realm in which even true and false remained unformed.

I looked up at what Miruka had written on the board. Axioms, inference rules, proofs, theorems... The most important concepts supporting mathematics. Could we really incorporate all of this into this 'miniature model' we were creating? Could we really end up doing math on mathematics?

I did the only thing I could do—sit back and follow where Miruka led.

5.3.5 *Axioms*

"We've defined formulas," Miruka said. "Now we need to define axioms."

Miruka turned and began writing on the board again.

"An axiom in our formal system H is a formula in one of these four patterns."

Axioms (Formal system H, definition 3)

An axiom is defined as a formula in x, y, z in any of the following forms:

Axiom P1 $((x) \vee (x)) \to (x)$

Axiom P2 $(x) \to ((x) \vee (y))$

Axiom P3 $((x) \vee (y)) \to ((y) \vee (x))$

Axiom P4 $((x) \to (y)) \to (((z) \vee (x)) \to ((z) \vee (y)))$

"These are also called axiom schemata," Miruka said. "Pick a schema, replace the x, y, z with formulas, and you've got an axiom."

She pushed her glasses up her nose and turned to me.

"Your turn," she said. "Is this an axiom?"

$$((A) \vee (A)) \to (A)$$

"Sure," I said. "P1 says that $((x) \vee (x)) \to (x)$ is an axiom, so just let x be A and you're done."

Miruka nodded.

"How about this one?"

$$(A) \rightarrow (A)$$

"Seems like it should work, but hang on... No, it doesn't. That's not an axiom."

"Why not?"

"It doesn't fit any of the patterns. P1 through P4 are the only definitions of axioms we've got, but none of them are in a pattern that fits."

"Good enough," Miruka said.

"Miruka..." Tetra said, her voice strained, "I have absolutely no idea whatsoever what we're doing."

"No?" Miruka said. "What don't you understand?"

"All of it! Well, no, I know we're talking about axioms, and I know you're saying that axioms have to be in certain patterns, but... but *why*? Where on earth did these patterns come from?"

"Let's set your mind at ease by looking a bit ahead."

5.3.6 Proof Theory

Miruka put down her chalk and sat on the edge of a desk, facing Tetra.

"We defined formulas as strings of characters as a step toward formally studying mathematics," she said. "Our next step is to formally define axioms, proofs, theorems and things like that. This is going to be easier than it should, because mathematicians who came before us—Hilbert in particular—have already found the axioms we need to develop formal systems. They've thought up a collection of formulas that can construct a formal system."

"So they proposed a set of axioms, deciding beforehand that they're true?" Tetra asked.

"No, there's still no true or false here."

"But how can we have axioms if there's no true or false?"

"Because in syntax, we think of axioms only in their relation to proofs. An axiom is a formula that you can use unconditionally when creating a proof. Just think of them as formulas that we take as theorems, even without a proof."

Tetra's cheek twitched, and she bit her thumbnail as she thought. After a time she spoke up.

"So something being true and something being proven are, like, different concepts?"

Miruka smiled.

"An excellent observation. You're right, but tracing through the philosophical implications of that statement would lead us away from the matter at hand."

Miruka stood and began slowly pacing.

"If you look at these axioms, you can probably tell that things can easily get so complex that humans would have trouble grasping what they say. $((A) \lor (A)) \to (A)$ isn't so bad, but something like $((A) \to (B)) \to (((\neg(A)) \lor (A)) \to ((\neg(A)) \lor (B)))$ isn't the kind of statement we want to work with. It's just too hard for a human to take in the overall structure. But don't forget, these axiom schemata are just strings of characters. It isn't hard for, say, a computer to read in something like that and tell you if what you have is an axiom. And we don't have to worry about computers getting bogged down with the meaning of any of this—they just mechanically read in the characters and act as axiom detectors."

"I'm sorry, one more question," Tetra said, raising her hand. "I'm still hung up on this idea of 'axioms.' Like, I see how P1 through P4 don't look like they give us $(A) \to (A)$. Fine, no problem. But going the other way is bugging me—I'm not sure why you're allowed to treat *anything* that you can plug into those templates as axioms."

"Hmm," Miruka said, placing a finger to her lips.

"Look at it this way," she said after a pause. "We're standing on the borderline between meaning and form. We're here because we want to formally study mathematics. To do that, we need to define the formulas we'll use to formally represent the assertions of mathematics. But if that's our goal, we need to formally define things like axioms, proofs, and theorems. Axioms are the stepping stones by which we arrive at proofs.

"So axioms in our formal system—let's call those 'formal axioms'—are formulas that will allow us to build 'formal proofs.' Those proofs in turn will allow us to create 'formal theorems.'"

Miruka went back to the board.

$$\begin{array}{rcl} \text{Mathematics} & \longleftrightarrow & \text{Formal systems} \\ \text{Propositions} & \longleftrightarrow & \text{Propositional formulas} \\ \text{Axioms} & \longleftrightarrow & \text{Formal axioms} \\ \text{Proofs} & \longleftrightarrow & \text{Formal proofs} \\ \text{Theorems} & \longleftrightarrow & \text{Formal theorems} \end{array}$$

"So in the end, all of mathematics can be represented as a formal system?" Tetra asked.

"That's a very deep question," Miruka said. "If you find a flower that has the shape of a rose, and the smell of a rose, and the color of a rose, it would be natural to call it a rose. So what we have to ask ourselves is, do formal systems have the color of mathematics? Do they share its scent and its form?"

Miruka cocked her head and smiled.

"But let's save that for another day."

5.3.7 Inference Rules

Miruka started pacing back and forth in front of the blackboard.

"So now we've defined formulas and axioms," she said. "We have axiom schemata P1 through P4 that give us the patterns axioms can come in, and we can stick formulas x, y, z into those patterns to create all the axioms we'd like. In other words, we can create infinitely many axioms, but only in a limited number of patterns. Axiom schemata don't allow us to create formulas in new patterns. To do that, we need to define an inference rule, something that will let us formally represent logical inferences."

Inference Rule (Formal system H, definition 4)

Inference rule MP For propositional formulas x, y, given x and $(x) \to (y)$ we can *infer* y.

"This particular inference rule is called *modus ponens*. That's what the MP stands for."

"Can you give us some examples of that?" asked Tetra.

"Sure. How about this?"

> Given formula A and
> formula $(A) \to (B)$,
> we can infer B by inference rule MP.

"Here's a slightly more complex example."

> Given formula $(A) \to (B)$ and
> formula $((A) \to (B)) \to ((\neg(C)) \vee (D))$,
> we can infer $(\neg(C)) \vee (D)$ by inference rule MP.

"See?" Miruka said. "You just follow the form."

Tetra raised her hand.

"Question?"

"So this *modus ponens* ... I shouldn't be reading that as saying that 'if x is true and x implies y, then we can infer y, right?"

"What do you think?"

"I don't think so, for the same reason as before—we're just using syntax to create a formal system, and there's no concept of true or false to apply here. This inference rule is just another form we're supposed to look at, without trying to assign meaning to it."

"An excellent answer to your own question."

"I never realized how hard it can be to read something without trying to apply meaning to it."

"You just have to get used to it. But don't worry, you will." Miruka smiled. "I know, it's impossible to think without meaning. And it's not like formal systems like these are created without purpose. There is indeed something going on in the background, a conscious attempt to create an interesting system. But interesting here means one that doesn't require normal human thought to proceed. It's one where operating formally, mechanically, gets you where you want to go."

"Leibniz's dream ... " I muttered.

"Is it possible to think without considering meaning?" Miruka mused. "More to the point, what *is* thought devoid of meaning, thought that even a machine can perform? How can we study formal mathematics as if we were machines?"

"Good question," I said. "How *do* we study formal mathematics?"

"Using mathematics, of course."

"So we're trying to find—"

"—the mathematics of mathematics."

5.3.8 Proofs and Theorems

"Let's see," Miruka said. "What do we have so far?"

She wrote a checklist on the board.

· We defined propositional formulas.

· We defined axioms.

· We defined an inference rule.

"Looks like we have everything we need to make a formal representation of proofs."

"That's it?" I said.

"Sure. Proofs are just inferences made on the basis of axioms, and we have all that. Now all we need to do is define how to represent them in our formal system H. And here's how we do it."

Proofs and theorems (Formal system H, definition 5)

Given a finite sequence of propositional formulas

$$a_1, \ a_2, \ a_3, \ \ldots, \ a_k, \ \ldots, \ a_n$$

such that for each formula a_k $(1 \leqslant k \leqslant n)$ either

(1) a_k is an axiom, or

(2) a_k can be inferred from formulas a_s and a_t, where s, t are natural numbers less than k,

then we call this sequence a *proof* of formula a_n. We furthermore call a formula a_n for which a proof exists a *theorem*.

"So this is actually a kind of double definition," Miruka said. "It gives us both proofs and theorems."

"So a proof is just a sequence of formulas?" Tetra asked.

"Not just any sequence," Miruka said. "There are rules regarding how you can line them up. Every time you add a formula to the sequence, it either has to be an axiom or something you can infer using two formulas you've already added to the sequence. Do you follow that?"

"I was with you up until the part about there being rules. After that . . . " Tetra gave an apologetic smile.

"Okay," Miruka said, slowing down a notch. "Say you want to line up some formulas to create a proof. There are two rules that limit what you can do." Miruka's chalk darted across the board.

1. You can add an axiom to the sequence at any time.

2. If you can use formulas that are already in the sequence to infer a new formula, you can add that inferred formula.

"When you follow these rules to create a sequence of formulas, then you can call the sequence a proof. Of course, when we talk about axioms here, we mean axioms that are valid in our formal system H, and inferences have to be made using the inference rule we've added to H. Get it?"

"So we can use axioms and formulas inferred from axioms?"

"Not quite. It's not just 'formulas inferred from axioms' that we can add. We can also add 'formulas inferred from formulas inferred from axioms.'"

"And formulas inferred from formulas inferred from formulas inferred from axioms," I added.

"And so on and so on, yes," Miruka said. "In other words, you can use any formula that's the result of a finite chain of inferences."

"Right!" Tetra said. "I think that's what I was trying to say."

"Anyway, you create a sequence of formulas that follows these two rules. When you've done that, you've created a proof, and you can call the last formula in the sequence a theorem. So what's a theorem, Tetra?"

"A formula that's backed up by axioms and inferred formulas!"

"Perfect," Miruka said. She put down the chalk and took a step toward us.

"Look what we've done here," she said. "We've defined formulas, axioms, inference rules, proofs, and theorems, the most fundamental aspects of mathematics. And we did it without writing a single number or geometric figure. No equations, no quadratic functions, no matrices. We just started from the bottom and worked our way up."

Dvořák's *Goin' Home* started playing over the intercom, the signal that we had to leave.

"That late already?" Miruka said, "They close the school up too early."

I looked out the window and saw it was already dark out.

Miruka raised a finger to catch my attention.

"Some homework for you," she said with a mischievous grin.

"Is $(A) \to (A)$ a theorem?"

5.4 Not Me, or Me

5.4.1 Good Intentions

That night I sat at my desk with every intention of studying for finals. I knew I should review my notes, at least, but somehow I couldn't find the motivation.

When it came to math I usually studied ahead in our textbook, even working out all the problems, so my classes were already like review work. I'd already studied beyond what we would cover that year, and I was getting 100s on every quiz. None of it was really a challenge anymore, not like the problems I found in other math books, or Mr. Muraki's cards, or the stuff that came up in our math talks.

So when I reached for my math notebook, my hand went for the one I used for my extra-curricular studies. I opened it and browsed for a while, smiling at the scribblings by Tetra, Miruka, and me.

When I came to a blank page, I listed the key points of our formal system H.

Formal system H: Summary

Formula F1 If x is a propositional variable, x is a formula.

Formula F2 If x is a formula, $\neg(x)$ is also a formula.

Formula F3 If x and y are formulas, $(x) \vee (y)$ is also a formula.

'Implies' symbol   Define $(x) \to (y)$ as $(\neg(x)) \vee (y)$.

Axiom P1 $((x) \vee (x)) \to (x)$.

Axiom P2 $(x) \to ((x) \vee (y))$.

Axiom P3 $((x) \vee (y)) \to ((y) \vee (x))$.

Axiom P4 $((x) \to (y)) \to (((z) \vee (x)) \to ((z) \vee (y)))$.

Inference rule MP   Given x and $(x) \to (y)$, we can infer y.

5.4.2   The Form of Form

I began to think about the homework Miruka had given me.

Problem 5-1 (Theorems in formal systems)

Is $(A) \to (A)$ a theorem in the formal system H?

I was sure the answer had to be 'yes,' but intuition counts for nothing in mathematics; I had to *prove* I was right.

But things were different this time. I was no stranger to doing proofs, but I was working under an unfamiliar set of rules. My proof had to be valid within this new formal system, a party to which my good friends contradiction and induction weren't invited. My toolbox had been pretty much emptied out, in fact, and I was left with only two techniques:

· Use the axioms of formal system H, and

· Apply the inference rule to create new formulas

*Just axioms and an inference rule... This feels like building
a house using just nails and a screwdriver.*

But no, I was sure there was a path from some axiom to $(A) \to$
(A); I just had to find it. There were limits to what I was allowed,
but that's always the case when doing math. Everything had just
been stripped down to a bare minimum.

A miniature model of mathematics, huh.

I sat there for a while, wondering where I should begin. I decided
I'd best follow the advice I was always giving Tetra, and start by
writing out some examples.

*I'm trying to work my way to $(A) \to (A)$, so at least I know
the variable will be A. Okay, so let x, y, z all be A, then.*

From axiom P1:

$$((A) \vee (A)) \to (A)$$

From axiom P2:

$$(A) \to ((A) \vee (A))$$

From axiom P3:

$$((A) \vee (A)) \to ((A) \vee (A))$$

From axiom P4:

$$((A) \to (A)) \to (((A) \vee (A)) \to ((A) \vee (A)))$$

I looked at what I'd written and found myself almost disap-
pointed.

Aw, c'mon Miruka. This is too easy for a problem from you.

P2 told me that $(A) \to ((A) \vee (A))$ was an axiom. In other words,
A implies $(A) \vee (A)$. Meanwhile P1 ensured that $((A) \vee (A)) \to (A)$
was an axiom, too. In other words, $(A) \vee (A)$ implies A. So all I had
to do was combine 'A implies $(A) \vee (A)$' and '$(A) \vee (A)$ implies A,'
and that would give me 'A implies A.'

I started to write this down, but something nagged at the back
of my mind. I paused, and ran this through my head once more.

... No, it doesn't!

I realized I hadn't been thinking syntactically—I was forcing my own interpretation of the word 'implies' on the → symbol, a definite no-no. If I was going to make inferences within this formal system, I was only allowed to use inference rule MP:

Given x and (x) → (y), we can infer y.

Okay, so how do I use this screwdriver to pound out some new theorems?

I thought. And I thought. And I thought some more.

I finally found myself in the zone, my head filling with variables and symbols coalescing into mathematical expressions. Among the infinitely many possible formulas, a few began to shine, studded with gleaming axioms.

Apply the inference rule to two axioms, and I get a theorem. Apply the inference rule to an axiom and a theorem, and I get another theorem. Apply the inference rule to two theorems, and I get a third... C'mon, what am I missing?

That's it!

(A) → (A) is a theorem, not an axiom, and the only way to create a new theorem is to apply the inference rule. So in the end what I was trying to do was find some way to arrange things so that the y in the inference rule—*modus ponens*, that was its name— became (A) → (A). That would let me infer (A) → (A) from x and (x) → ((A) → (A)). There was just no other way to get where I wanted to go.

A fine realization, but one that left a big question: what formula for x would let me do that?

5.4.3   The Meaning of Meaning

I sat for a while, staring at the letters on the page before me. I added a new line to focus on.

I can infer (A) → (A) from x and (x) → ((A) → (A))

The goal before me was finding some value for x that made this work. All I had to work with so far were a handful of theorems I'd created from the axioms, so I started plugging those in to see what would happen.

Might as well start at the top of the list, letting x *be* A *to get* $((A) \vee (A)) \to (A)$.

I can infer $(A) \to (A)$ from $((A) \vee (A)) \to (A)$ and $(((A) \vee (A)) \to (A)) \to ((A) \to (A))$

All well and good, but for this to be helpful I'd have to somehow create $(((A) \vee (A)) \to (A)) \to ((A) \to (A))$.

Can a mess like this really be straight from the axioms? Sure would be nice if it was...

I compared the pattern of each axiom P1 through P4, looking for the one that was most similar.

P1, I guess? That one says $((x) \vee (x)) \to (x)$. *If I plug* $(A) \to (A)$ *into this* x, *I get... let's see...* $(((A) \to (A)) \vee ((A) \to (A))) \to ((A) \to (A))$.

I rapidly filled the page with letters, parentheses, arrows, and wedges. As I did so, I developed a new appreciation for what Tetra had said about it being hard to do math without thinking about meaning.

5.4.4 Implies 'Implies'?

After a while I realized that I'd only been playing with P1 and P2, despite having four axioms at my disposal. I decided to give P4 a shot and see what fell out of that.

Axiom P4:
$$((x) \to (y)) \to (((z) \vee (x)) \to ((z) \vee (y)))$$

Just looking at it was enough to tell me this wouldn't work— *modus ponens* gave what was on the right side of the $\to$ operator. It said that you can infer ♠ from ♡ and $(\heartsuit) \to (\spadesuit)$. But P4 looked like this.

$$((x) \to (y)) \to (((z) \vee (x)) \to (\underset{\sim}{(z) \vee (y)}))$$

So P4 would in the end just give $\underset{\sim}{(z) \vee (y)}$, not the $(A) \to (A)$ I was after.

So where can I use P4?

I thought back on what Miruka had done in her lecture that day. She'd defined formulas, then axioms and inference rules, and finally proofs and theorems.

I found myself thinking syntax was interesting in a way, but kind of a pain to deal with. My mind drifted back to Leibniz's dream of mechanical thought, and how this kind of math might be best left to computers. I thought about how I'd created truth tables with Yuri, using semantics without even realizing it. Even Yuri had gotten hung up on what 'implies' means, despite her skill at logic. Once you get used to it, mechanically thinking of 'A implies B' as meaning 'not A, or B' starts to come naturally, though. That's just the form of 'implies.'

The form of 'implies'...

...Hm?

My heart skipped a beat. I was on to something.

We had a definition for the 'implies' symbol $\rightarrow$ in formal system H. I even remembered Miruka up at the board defining it. "Let's make formulas in our formal system H a little easier to read," she'd said.

'Implies' symbol   Define $(x) \rightarrow (y)$ as $(\neg(x)) \lor (y)$.

That's it!

Instead of using *modus ponens* to infer $(A) \rightarrow (A)$, I could infer $(\neg(A)) \lor (A)$ instead. Axiom P4 just might come in handy after all!

Axiom P4:

$$((x) \rightarrow (y)) \rightarrow (((z) \lor (x)) \rightarrow ((z) \lor (y)))$$

Yes! I can just let z be $\neg(A)$, and let y be A!

$$((x) \rightarrow (A)) \rightarrow (((\neg(A)) \lor (x)) \rightarrow ((\neg(A)) \lor (A)))$$

So far so good! Now I just need to think about what to stick into x...

$$((x) \rightarrow (A)) \rightarrow (((\neg(A)) \lor (x)) \rightarrow ((\neg(A)) \lor (A)))$$

I stared at what I had underlined, but the answer came much faster this time. I just had to let x be $(A) \vee (A)$. Then $(\underline{x}) \to (A)$ would become this.

$$((\underline{A) \vee (A)}) \to (A)$$

That's the form of axiom P1!

With this, $(\neg(A)) \vee (\underline{x})$ becomes $(\neg(A)) \vee (\underline{(A) \vee (A)})$, and I could write $(\neg(A)) \vee ((A) \vee (A))$ using the $\to$ symbol.

$$(A) \to ((A) \vee (A))$$

Axiom P2! Everything is falling into place now!

I went back through the various formulas I'd scattered about, and pulled them together into a cleaner proof.

L1. Let x in axiom P1 be A.
   $((A) \vee (A)) \to (A)$

L2. In axiom P4, let x be $(A) \vee (A)$, let y be A, and let z be $\neg(A)$.
   $(((A) \vee (A)) \to (A)) \to (((\neg(A)) \vee ((A) \vee (A))) \to ((\neg(A)) \vee (A)))$

L3. In axiom P2, let x and y both be A.
   $(A) \to ((A) \vee (A))$

L4. Apply the inference rule MP to formulas L1 and L2.
   $((\neg(A)) \vee ((A) \vee (A))) \to ((\neg(A)) \vee (A))$
   We can rewrite as follows:
   $((A) \to ((A) \vee (A))) \to ((A) \to (A))$

L5. Apply the inference rule MP to formulas L3 and L4.
   $(A) \to (A)$

Done!

I finally had a proof that $(A) \to (A)$ in formal system H. I'd shown that in this system, $(A) \to (A)$ was a theorem!

Answer 5-1 (Theorems in formal systems)

$(A) \to (A)$ is a theorem in formal system H, as follows:

L1. $((A) \vee (A)) \to (A)$

L2. $(((A) \vee (A)) \to (A)) \to (((\neg(A)) \vee ((A) \vee (A))) \to ((\neg(A)) \vee (A)))$

L3. $(A) \to ((A) \vee (A))$

L4. $((\neg(A)) \vee ((A) \vee (A))) \to ((\neg(A)) \vee (A))$
This can also be represented as
$((A) \to ((A) \vee (A))) \to ((A) \to (A))$

L5. $(A) \to (A)$ □

P1. $((x) \vee (x)) \to (x)$ P2. $(x) \to ((x) \vee (y))$

L1. $((A) \vee (A)) \to (A)$ L2. $(((A) \vee (A)) \to (A)) \to (((\neg(A)) \vee ((A) \vee (A))) \to ((\neg(A)) \vee (A)))$

MP

P2. $(x) \to ((x) \vee (y))$ L4. $((\neg(A)) \vee ((A) \vee (A))) \to ((\neg(A)) \vee (A))$

L3. $(A) \to ((A) \vee (A))$ $((A) \to ((A) \vee (A))) \to ((A) \to (A))$

MP

L5. $(A) \to (A)$

Flow of the proof of $(A) \to (A)$

5.4.5 Asked Out

The next morning, the phone rang as I was getting ready for school. When I answered, a familiar but unexpected voice wished me good morning.

"*Miruka*? Er . . . hey, what's up?"

"Lemme talk to your mom."

"My *mom*? Uh, sure. Hang on."

I had no idea what business Miruka could have with my mother, but what else could I say?

"Mom! Phone!"

My mother appeared from the kitchen, wiping her hands on her apron.

"Who is it?"

"Miruka."

"For *me*? How curious... Hello? Yes, dear, how can I help you?"

The situation was too uncomfortable for me to stand there listening in, so I wandered off. I couldn't hear exactly what my mother was saying, but her laughs and ebullient tone of voice painted a good mental image of the scene. The merry conversation ended with a click, and I rushed back into the room.

"You hung up?" I said. "She didn't want to talk to me at all? What did she want?"

"To go out on a date," my mother said, grinning.

"Miruka wants to take you out on a *date*?"

"Not me, silly. You."

"Wha—?"

At that point I was only sure of one thing—the day couldn't get any weirder.

Trying to swim, you imitate what other people do with their hands and feet to keep their heads above water, and, finally, you learn to swim by practicing swimming. Trying to solve problems, you have to observe and to imitate what other people do when solving problems and, finally, you learn to do problems by doing them.

GEORGE POLYA
How to Solve It: A New Aspect of Mathematical Method

CHAPTER 6

Epsilon Delta

Remembering the words the captain of the robbers used to cause the door to open and shut, he wished to try if his pronouncing them would have the same effect. Accordingly he went among the shrubs, and, revealing the door concealed behind them, stood before it, and said, "Open, Simsim!" Whereupon the door instantly flew wide open.

Stories from the Thousand and One Nights
TRANS. BY EDWARD WILLIAM LANE

6.1 LIMITS OF SEQUENCES

6.1.1 Crash Course

Entering the library after school one day I nearly collided with Tetra, who was just on her way out.

"Whoa!"

"Sorry!" Tetra said. "I was off to look for Miruka."

"She's not here? She just said 'later' and left class as soon as the bell rang. I figured this is where she'd be."

"Aw, darn. I wanted to learn how to use equations to define limits, like she was talking about the other day."

Tetra looked up at me with those big brown eyes, and I had no doubt how I'd be spending the afternoon.

"I'm pretty sure I know what she meant, if you don't mind me showing you instead."

"That would be great!"

"How about we move to the lecture hall, though. It'll be easier if we can use the big board they have there."

"You bet!"

6.1.2 In the Lecture Hall

The lecture hall was a special classroom that mainly got used for science labs and demonstrations. The room was tiered, curving around a podium at the bottom.

The room was pleasantly cool, and the faint odor of chemistry experiments gone awry hung in the air. Tetra and I went behind the podium and stood facing the blackboard.

"Okay then," I said, picking up a piece of chalk. "You represent the limit of a sequence like this."

$$\lim_{n \to \infty} a_n = A$$

"In words, this means the sequence $\langle a_n \rangle$ converges as n approaches infinity, and that the limit it converges to is A."

"So n is some variable," Tetra said, "and the bigger n gets, the closer a_n gets to some number A, right?"

"Yeah, that's a good way to think about it at a conceptual level," I said, "but words like 'bigger' and 'closer' are kinda vague, aren't they? In a mathematical sense, at least. What we'd really like to do is get away from words like that, so we can really nail down what's going on."

"And that's where the equations come in?"

"Not an equation, exactly. A logic statement. We want to say that if a sequence $\langle a_n \rangle$ has a limit A, then this statement is true."

I turned and wrote on the board.

$$\forall \epsilon > 0 \; \exists N \; \forall n \left[n > N \Rightarrow |a_n - A| < \epsilon \right]$$

"Whoa," Tetra said. "What's this?"

"Our final goal," I said. "We want to understand everything about it. Because if you understand this statement, then you understand limits."

"Question!" Tetra said, raising her hand. "Miruka said something about epsilons and deltas. Where do they come in?"

"Defining limits this way is called using the '(ϵ, δ)-definition of a limit.' It comes from the Greek letters."

α	alpha
β	beta
γ	gamma
δ	**delta**
ϵ	**epsilon**
$\vdots$	$\vdots$

"How do Greek letters make the definition?"

"Epsilon and delta are two variables often used in the definition. They play a vital role, so it's named after them."

"Why didn't you use both of them? I only see epsilons."

$$\forall \textcircled{\epsilon} > 0 \ \exists N \ \forall n \ \left[n > N \Rightarrow |a_n - A| < \textcircled{\epsilon} \right]$$

"When you're talking about the limit of a *sequence*, it's common to use (ϵ, N). Using (ϵ, δ) is normally used for limits of a *function*. We'll talk about that later."

"Do you have to use Greek letters?"

"Not if you don't want to. The math's the same in either case."

Limit of a sequence (ϵ-N representation)

$$\lim_{n \to \infty} a_n = A$$
$$\Updownarrow$$
$$\forall \epsilon > 0 \ \exists N \ \forall n \ \left[n > N \Rightarrow |a_n - A| < \epsilon \right]$$

"Okay. But I'm still not sure what's wrong with just using the bigger–closer definition. Seems a lot clearer than this alphabet soup!"

"Because it doesn't hold up when you try to examine limits precisely."

"What do you mean, exactly?"

"Well, when you talk about a_n getting 'closer and closer' to A, you're probably thinking about something like this."

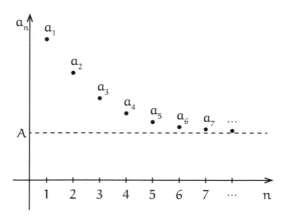

"Yeah! That's what's going on here, right?"

"That's one scenario, but other situations are possible too. What if a_n actually becomes *equal to* A? Would you still say it's getting closer?"

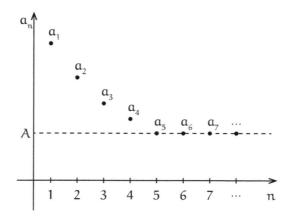

"Hmm, that's a tricky one. Is 'equals' the same as 'close'? Maybe, maybe not."

Tetra squinted at her thumb and forefinger as she slowly brought them closer and closer together.

"See? That's a perfect example of why 'closer' is a vague word. There's other situations, too. What if a_n gets closer to A, then farther? Or what if it jumps back and forth over A?"

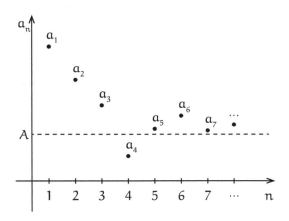

"Ah, right. I see."

"The more you think about it, the more problems there are with that kind of definition. It isn't precise, so different people can interpret it in different ways. So let's get rid of all the words and see how math is a better language to describe this with."

"Sounds good!" Tetra said, beaming.

6.1.3 Understanding Complex Statements

"Our first step is figuring out what all this means."

I pointed to the long statement on the board.

$$\forall \epsilon > 0 \; \exists N \; \forall n \; \left[n > N \Rightarrow |a_n - A| < \epsilon \right]$$

Tetra said nothing.

"Yeah, I know," I said. "It's a bit complex."

"It sure is! I'm trying very hard not to freak out here."

Tetra put a hand to her chest. She'd never been a big fan of using multiple variables, and this was likely an overdose.

"No need to freak out," I said. "There's nothing in here you don't understand, we just have to take it a piece at a time. You can master this definition by divide and conquer, splitting it up into smaller chunks and thinking them through."

"I'll give it my best shot," Tetra said with a firm nod.

"First, let's look at how this statement is structured."

$$\forall \epsilon > 0 \quad \exists N \quad \forall n \left[\; n > N \; \Rightarrow \; |a_n - A| < \epsilon \; \right]$$

"There's a couple of $\overset{\text{for all}}{\forall}$ and $\overset{\text{exists}}{\exists}$ symbols in here, right? Let's make their scope a little bit clearer, by adding some more brackets."

$$\forall \epsilon > 0 \left[\; \exists N \left[\; \forall n \left[\; n > N \; \Rightarrow \; |a_n - A| < \epsilon \; \right] \; \right] \; \right]$$

"And now, let's read through these in turn. It says that for any positive number ϵ—"

$$\underset{\sim}{\forall \epsilon > 0} \left[\hspace{8cm} \right]$$

"—there exists some natural number N—"

$$\forall \epsilon > 0 \left[\; \underset{\sim}{\exists N} \left[\hspace{6cm} \right] \; \right]$$

"—such that something is true for any natural number n."

$$\forall \epsilon > 0 \left[\; \exists N \left[\; \underset{\sim}{\forall n} \left[\hspace{4cm} \right] \; \right] \; \right]$$

"Oh, okay!" Tetra said. "So there's three sets of brackets setting things off."

"Right. So let's fill in that innermost set of brackets."

$$\forall \epsilon > 0 \left[\; \exists N \left[\; \forall n \left[\; n > N \; \Rightarrow \; |a_n - A| < \epsilon \; \right] \; \right] \; \right]$$

"I think I get the structure," Tetra said, "but I need to write this down in words, or I'm pretty sure I'll forget."

"No problem," I said. "I'll dictate."

For every positive number ϵ,

there exists some natural number N

such that for every natural number n,

$n > N \Rightarrow |a_n - A| < \epsilon$ is true.

Tetra nodded. "Got it. But that's still quite a mouthful."

"Hmm... Maybe in this case using more words would help."

For every positive number ϵ,

if you choose an appropriate natural number N
for that value of ϵ,

then for every natural number n,

we can set things up so that $n > N \Rightarrow |a_n - A| <$
ϵ will be true.

"Okay, I'm not totally there yet, but at least the impending freak-out feeling has subsided."

"You'll feel even better after you've written this out on your own a few times, thinking it through as you do."

"One thing I noticed... You kept saying there exists *a natural number* here and there, but that's not in the statement. Doesn't it need to be?"

"I just left that out to simplify things. You can write it if you want to, of course. Then the statement would look like this."

$$\forall \epsilon > 0 \left[\quad \underset{\sim\sim\sim}{\exists N \in \mathbb{N}} \left[\quad \underset{\sim\sim\sim}{\forall n \in \mathbb{N}} \left[\quad n > N \quad \Rightarrow \quad |a_n - A| < \epsilon \quad \right] \quad \right] \quad \right]$$

"This makes things even longer," I said, "but the principle remains the same—when reading complex math, divide and conquer."

6.1.4   Reading Absolute Values

"It's still a lot of variables," Tetra said.

"It's probably not as bad as you think," I said. "Count them and see."

$$\forall \epsilon > 0 \left[\exists N \left[\forall n \left[n > N \Rightarrow |a_n - A| < \epsilon \right] \right] \right]$$

"Let's see, there's ϵ, big N and little n, A and a_n, and... uh... Hey, there's only five. Wow, I thought there'd be a lot more."

I shrugged.

"It just looks that way because some of them get reused. So tell me, what do A and a_n mean? Do you remember what they represent?"

"Umm... A is the limit, right? And a_n is the number that's getting really close. The number that we're paying attention to."

"Good. Put a little more precisely, a_n is the nth element in the sequence $\langle a_n \rangle$. So a_1 is the first element, and a_{123} is the 123rd element, and so on."

"Yep, I'm good."

A	The limit of sequence $\langle a_n \rangle$
a_n	The nth element in sequence $\langle a_n \rangle$

"Okay, so what does $\left| a_n - A \right| < \epsilon$ tell us?" I asked.

"That the absolute value of $a_n - A$ is smaller than ϵ," Tetra said, nodding.

"And do you see what the absolute value of $a_n - A$ stands for?"

"Uh... honestly, no."

"It's the *distance* between point A and point a_n on the number line."

"The distance?"

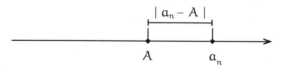

$\left| a_n - A \right|$ is the distance between two points

"That's right. And since it's an absolute value, it doesn't matter if point a_n is to the left or the right of point A."

"Ah, so we don't care *where* it is, just *how far away* it is."

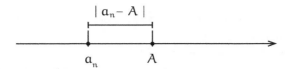

$|a_n - A|$ **gives the distance, even when a_n is to the left of A**

"Exactly," I said. "So what does it mean to say that that distance is smaller than ϵ?"

"Umm... that point a_n doesn't get very far from point A?"

" 'Not very'? Can you be a little more precise?"

"Oh, point a_n can't get as far as ϵ from point A!"

"There ya go. Point a_n can only move in the range of the thick line here."

The distance between point a_n and point A is less than ϵ

"Okay, so it can only go as far right as $A + \epsilon$ and as far left as $A - \epsilon$."

"Almost perfect—more precisely, a_n has to remain *within* the space between $A - \epsilon$ and $A + \epsilon$."

"Oh, I see! It's not allowed to step on the white dots."

"There's a special name for this range that a_n is allowed to move around in: the 'ϵ-neighborhood.' "

"How cute!"

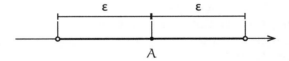

The ϵ-neigborhood of A

"Uh, I guess," I said. "Anyway, this point a_n will be somewhere in the ϵ-neighborhood of A."

"Wait a sec, it can wander away from A, so long as it doesn't leave the neighborhood?"

"Up to a distance ϵ away, right."

"But that doesn't feel very much like getting arbitrarily close to A..."

"That's an excellent point, but remember that since this is supposed to be true for every positive ϵ, you could choose a new, smaller ϵ to make things get closer to A. But anyway, right now just be sure you understand these two things."

I wrote two bullet points on the board.

· Point a_n is within distance ϵ of A.

· Point a_n is in the ϵ-neighborhood of A.

"Okay, got it," Tetra said. She tilted her head and smiled. "It's funny. Y'know how just a few minutes ago I said that the absolute value of $a_n - A$ has to be less than ϵ? Well I knew what I was saying was right, but I wasn't sure exactly what it *meant*. But this graph you drew makes it totally clear, and calling the place that a_n has to be in its neighborhood is even better. It's funny how representing the same thing in different ways totally changes how you understand it."

"I agree. Presentation is everything."

Tetra gasped and covered her mouth.

"What's wrong?" I asked.

"Absolute values reminded me of something—last year you made me write a definition of absolute values right here in this very room, the first time you gave me a math lesson. I tripped on the stairs here and almost knocked you over. I was *so* embarrassed."

"I remember. You still crash into things, but at least you've made a lot of progress in math. Your hard work is paying off."

"Thanks to you!"

6.1.5 Reading 'Implies'

"Moving on," I said. "You see what this means now, right?"

$$n > N \Rightarrow |a_n - A| < \epsilon$$

"Sure!" Tetra said. "Er, maybe. I still don't know what N is."

"That's okay. Just read the statement as-is."

"Then I guess it means that if small n is bigger than big N, then the distance between a_n and A is less than ϵ."

"Perfect. Can you write that on the board, but using the neighborhood idea this time?"

"Something like this?"

> If n is bigger than N, then a_n is in the ϵ-neighborhood of A.

"Well done."

"So does this mean that when n is big, a_n is closer to A?"

"Yeah, but let's try to express that a little more quantitatively. If someone asks 'when n is big, how big is it?' you can say 'bigger than N.' If they ask 'how close does a_n have to be to A for it to be considered close?' you can say 'it has to be in A's ϵ-neighborhood.' In other words, this expression $n > N \Rightarrow |a_n - A| < \epsilon$ is telling us something about the relationship between the size of n and the distance between A and a_n."

"Okay, I think I get that. Hang on, though, I've got to write some stuff down before I forget everything we've been talking about."

Tetra pulled a notebook out of her bag, and I watched as she made some notes.

- Break complex statements up into smaller pieces.

- Don't freak out when Greek letters show up.

- Think about what variables mean.

- Think about what 'absolute value' means.

- Represent things as graphs.

- Think about what inequalities imply.

"Good summary," I said. "But each of those is pretty much common sense, right?"

"Yeah, but they're worth repeating. I try to take everything in at once, and I panic when I can't. So I especially need to keep reminding myself to chop things up into smaller bits."

Tetra pantomimed cutting something up, though it looked more like she was chopping vegetables than mathematical expressions.

6.1.6 Reading 'For All' and 'Some'

"Okay," I said. "Let's try taking on the whole thing."

"You bet!" Tetra said, clutching her fists.

$$\forall \epsilon > 0 \left[\exists N \left[\forall n \left[n > N \Rightarrow |a_n| - A < \epsilon \right] \right] \right]$$

"I'm going to write this all out in plain words, so we can take our time and comb through it."

> For every positive number ϵ,
>
> if you choose an appropriate natural number N for that value of ϵ,
>
> then for every natural number n,
>
> we can set things up so that if N is smaller than n, a_n will be in the ϵ-neighborhood of A.

"Now," I said, putting down the chalk, "read that through a few times, taking as much time as you like, and tell me what you find."

I stood to the side, watching Tetra. She put a hand to her mouth and stared in silence at what I'd written. After a while, she nodded slowly.

"I think I'm seeing something. Correct me if I'm wrong, but is this saying that if you make ϵ really, really small, you can squeeze a_n into a really tiny neighborhood?"

"Excellent! It does indeed say that."

"Well I'm glad I got that right, at least. But I still don't know what this N is... What's it supposed to represent?"

"Good question. N is a number that represents how big you have to make n to keep a_n in A's ϵ-neighborhood. We don't care about n's that aren't bigger than N. But when we have an n that fits this condition, all a_n's will be in the neighborhood of A."

"Um..."

"Think of it this way. You have a sequence $\langle a_n \rangle$ that converges to A. A challenger appears, bearing a *really* small epsilon. He says, 'You think that sequence gets really close to A, huh? Well can it get so small that it fits into *this* teeny, tiny ϵ-neighborhood?' You say, 'Sure. If we ignore the first N terms in the sequence, every one of my a_n's will fit in there.'"

"Oh, wow..." Tetra said.

"Don't forget the order that ϵ and N come in." I pointed at the equation on the board.

$$\forall \underset{\sim}{\epsilon \geq 0} \left[\exists \underset{\sim}{N} \left[\forall n \left[n > N \Rightarrow |a_n - A| < \epsilon \right] \right] \right]$$

"This says that ϵ gets chosen first, and then we get to choose an N after seeing what the ϵ is. So our response to the challenge of a really small ϵ is just choosing a really big N. That lets us lop the first N terms off of our sequence, and we'll *still* have infinitely many terms happily squeezed into the ϵ-neighborhood. That's what (ϵ, N) asserts, and that's what it means for a sequence to converge."

"Okay, I think I'm getting this," Tetra said. "It doesn't matter how small the ϵ-neighborhood is, we just have to use ϵ to decide on an N, which tells us how many of our first terms to ignore so that all the rest fit into the ϵ-neighborhood. Got it."

"Good. Also, remember that ϵ itself is some finite number. We aren't saying that it's 'infinitely small' or anything. Just that no matter how small it gets, we can deal with it. That's key, because it lets us define limits without having to use tricky concepts like infinitesimal values."

"So why do we have to use this N? All we want to say is that if we make n bigger, it's possible to squeeze a_n's into this ϵ-neighborhood around A. I'm not sure where this new variable N comes into the picture."

"That's exactly why we have this $\exists N$ here."

"To do... what?"

"To get the 'it's possible to...' part into the expression. The $\exists$ symbol is saying 'there exists some number that satisfies this condition.'"

"So we're expressing the possibility of a thing through the existence of some number?"

"Exactly. How about we get a taste of the power of (ϵ, N)?"

"Huh?"

"Say that $a_n \to A$ as $n \to \infty$. In that case, is it okay to have an a_k where $a_k = A$?"

"Can it hit A right on the nose, you mean? Sure, that shouldn't be a problem. The only thing that's important is that a_k stays within the ϵ-neighborhood of A, and if $a_k = A$ it's absolutely in the neighborhood. It's not even leaving the house!"

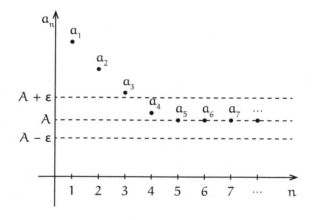

"That's right. What about the case where a_n gets closer to A, but then moves away from it?"

"Yeah, I'd think that would be okay, too. So long as it doesn't leave the neighborhood. But I don't think it's allowed to just sit in one place, always the same distance from A. In the long run it has to keep getting closer, right? If it didn't, it seems like after some point there would be an ϵ-neighborhood that the sequence couldn't squeeze into, no matter how many terms you chopped off. Umm... am I saying that right? I think I've got the right mental picture, but it's hard to put in words. What I'm trying to say is that you can't have this."

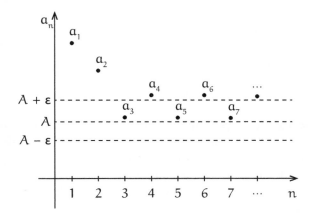

Tetra grimaced.

"It's hard to put such tricky conditions into words, isn't it," she said. "So yeah, I guess using variables like N here is the best way to be sure you're understood."

"I'm glad to see you're starting to feel that for yourself. Once you've wrestled with a few tough expressions like this, they won't feel as complex as they used to. Then one day you'll see that they aren't scary at all, and you can focus on what they say, not what they look like."

"*That* will be a good day!"

"Now that you're getting used to this one, let's get rid of these extra brackets."

$$\forall \epsilon > 0 \; \exists N \; \forall n \left[n > N \Rightarrow |a_n - A| < \epsilon \right]$$

"It may not be scary anymore," Tetra said, "but it's still pretty intimidating! I'm getting there, though!"

6.2 THE LIMITS OF FUNCTIONS

6.2.1 *Epsilon–delta*

"We've been talking about the limits of sequences," I said, "but let's move on to functions."

"Okay!" Tetra said.

"With sequences we were using (ϵ, N), but with functions we're going to use (ϵ, δ). We'll look at this the same way we did sequences. First off, the limit of a function is written like this."

$$\lim_{x \to a} f(x) = A$$

"In words this says, 'function $f(x)$ converges as x approaches a, and the value of its limit is A.' A simplified version that a lot of people are first introduced to says, 'as x gets arbitrarily close to a, the value of function $f(x)$ gets arbitrarily close to A.' In any case, we can define function $f(x)$ having a limit A as the real number x approaches a using an expression like this."

$$\forall \epsilon > 0 \; \exists \delta > 0 \forall x \left[0 < |x - a| < \delta \Rightarrow |f(x) - A| < \epsilon \right]$$

The limit of a function ((ϵ, δ) representation)

$$\lim_{x \to a} f(x) \quad = \quad A$$

$$\Updownarrow$$

$$\forall \epsilon > 0 \; \exists \delta > 0 \; \forall x \left[0 < |x - a| < \delta \quad \Rightarrow \quad |f(x) - A| < \epsilon \right]$$

"See if you can add the brackets in, like I did before," I said. "Something like this?"

$$\forall \epsilon > 0 \left[\; \exists \delta > 0 \left[\; \forall x \left[\; 0 < |x - a| < \delta \; \Rightarrow \; |f(x) - A| < \epsilon \; \right] \; \right] \; \right]$$

"Good job," I said. "Now can you read all that?"

"It's pretty much like before, right? So starting from the outside it's saying that for any positive number ϵ ...

$$\underset{\sim}{\forall \epsilon > 0} \left[\right]$$

" ... there exists some positive number δ ...

$$\forall \epsilon > 0 \left[\quad \underset{\sim\sim}{\exists \delta \geqslant 0} \left[\qquad\qquad\qquad\qquad\qquad\qquad \right] \quad \right]$$

" ... such that for any value of x ...

$$\forall \epsilon > 0 \left[\quad \exists \delta > 0 \left[\quad \underset{\sim}{\forall x} \left[\qquad\qquad\qquad \right] \quad \right] \quad \right]$$

" ... if $|x - a|$ is positive and less than δ, then $|f(x) - A|$ is less than ϵ."

$$\forall \epsilon > 0 \left[\quad \exists \delta > 0 \left[\quad \forall x \left[\quad 0 < |x - a| < \delta \;\Rightarrow\; |f(x) - A| < \epsilon \quad \right] \quad \right] \quad \right]$$

"Excellent! And what does this mean?"

I pointed at the left part of the innermost expression.

$$0 < |x - a| < \delta$$

"Umm ... the absolute values were indicating distances, right? So this says that a and x aren't the same number, and that the distance between them is less than δ. I think."

"You're absolutely right, but remember how we were using that word neighborhood?"

"Oh, right! So x is in the—huh?" Tetra looked back and forth between this expression and the one we worked on when talking about the limits of a sequence. "Wait, things have changed."

"Yep, we're not in the ϵ-neighborhood anymore."

"We moved to the δ-neighborhood!"

"Right. So what does it mean to say $0 < |x - a| < \delta$?"

"That x is in the δ-neighborhood of a, right? And that x isn't sitting right on top of a?"

"Because $0 < |x - a|$, right. So putting it all together, this innermost part is saying, 'if there's some x that isn't equal to a in the δ-neighborhood, then $f(x)$ is in the ϵ-neighborhood of A."

"There's *two* neighborhoods this time!"

"Yep. When you're dealing with the limit of a function and you're challenged with a small epsilon, you come back with a delta, saying that if you place x in the δ-neighborhood of a, then $f(x)$ will be somewhere in the ϵ-neighborhood of A. It's the existence of the appropriate δ that makes this work."

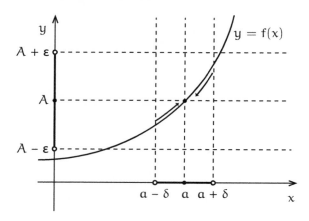

6.2.2   *The Meaning of* (ϵ, δ)

I glanced at the window and saw it was getting dark outside.

"We better be getting home soon," I said. "But before that, let's be sure you're getting this whole (ϵ, δ) thing. Do you remember why we wanted to use it in the first place?"

"Uh, gimme a second... Oh, yeah. It's because we want to avoid phrases like 'arbitrarily close' to give a more precise definition of limits."

"Right. And what did we use in place of 'arbitrarily close'?"

"Erm, that's the 'for any positive number ϵ, no matter how small' part, right? We're saying that no matter what ϵ somebody chooses, we can find a δ that will make this all work."

"Right, and that's the heart of all this. When you talk about the existence of a limit in (ϵ, δ) terms, the most important thing is that we're guaranteeing that we can find an appropriate δ, regardless of what value for ϵ is thrown at us. You don't get that with the 'arbitrarily close' approach."

6.3 Finals

6.3.1 Unranked

Tetra and I left the lecture hall and headed down a hallway that circled the courtyard, leading toward the exit. Partway down the hall, we noticed a group of students standing near the principal's office, looking up at the wall..

"Oh, they must have posted the ranks," Tetra said.

"I guess so."

At my school, they posted the names of students scoring in the top 10% of finals for each subject. Everyone would be listed in order of highest score on a big sheet of paper, displayed for all to see. We called this 'the ranking,' and if your name was on the list you were 'ranked.' Of course I headed straight for the math rankings for my grade, confident that I'd see my name listed, as usual.

Miruka was ranked, of course, as was Kaito and all the others I expected to see. The math rankings didn't change much from semester to semester, and all of the usual suspects seemed to be accounted for ... except for one.

"I don't see your name," Tetra said.

"Yeah, me neither."

"Off day?"

"It happens," I said, but inside I was screaming *no way!*

"Oh, wow! Look! Look!" Tetra was gesturing frantically at the math rankings for her grade, beaming. "My first math ranking!"

I acknowledged Tetra's achievement with a grim nod, then glanced at the list next to where she was pointing—English. Sure enough, her name was there, too.

"You double-ranked," I said. "Congratulations."

"Thanks!" Tetra said, flushed with excitement and taking no note of my wooden tone.

I turned away from the lists, my face as red as Tetra's.

6.3.2 Sound of Calm, Voice of Silence

Tetra and I left the school grounds and walked side-by-side toward the station. I reflected on my math final, about how I hadn't paid much heed to the weird feeling I'd had after finishing it. The section on integrals had given me particular trouble. They had been fairly straightforward, all solvable using the standard formulas, but there were so many I'd had to rush. I'd figured everyone else would be in the same boat, and that I'd be able to ride the curve to a top score like I usually do.

How stupid of me. How truly stupid.

"So about that definition of limits..." Tetra said, business as usual.

"What about it?"

"I see the difference between the (ϵ, δ) definition and the 'arbitrarily close' one, but something's still bugging me."

"Yeah?" I said, finding the idea of talking math just then arbitrarily close to annoying.

"Well, it's neat that we got through that monster expression and all, but... what's the point? How do we use it?"

"Well..." *It's neat that you want to have this conversation and all, but what's the point?* "You can use that definition of limits to define derivatives and integrals, for one thing. Continuity, too, if you know what that means."

"Continuity, like things coming one after another?"

"That's the dictionary definition."

"Meaning I shouldn't use it?"

"Not when you're doing math. You lose the precision that math is all about."

"Huh, okay," Tetra said. She continued on, half talking to herself as she often did.

"It's funny how logic feels so unlike working with equations and doing calculations and stuff like that. Remember how I said that doing math with integers felt 'squeaky'? Well this is different again. Logic problems feel... quieter, calmer. That's it, they sound like silence! You know how when you think things are really quiet, but when you pay attention and listen close, you can actually hear all kinds of little noises? Following a logic statement kinda reminds me of that. I wonder if Ay-Ay would enjoy this stuff... Anyway, it's kinda cool how different areas of mathematics have their own feel to them. Where do you think that comes from?"

I couldn't manage a response.

"You okay?" Tetra asked.

"Fine."

She must have finally caught on to my mood. She looked down at the road, and we didn't speak again until we reached the station.

"I'm going to head to the bookstore, so..." Tetra began.

I simply nodded. Tetra tilted her head, and flicked her fingers *one, one, two, three*—the first part of our Fibonacci Sign.

"See ya," I said.

I turned toward the ticket gate and walked away.

6.4 CONTINUITY

6.4.1 In the Library

The problem with wanting to revel in a blue funk is that a new day always dawns. I plodded through my classes the next day, my misery fading into embarrassment as I realized how childish I was being.

Miruka walked by my desk when our last class was done, but I was still gathering my things.

"Slowpoke," she said, making me wince. "See ya in the library."

When I finally made it there, Miruka and Tetra were already deep in conversation.

"So let's look at an equation that gives a specific definition of continuity," Miruka said.

"Do you just have stuff like that memorized?" Tetra asked.

"It isn't memorization once you have a good grasp on what it means. Here, look at this."

Definition of continuity (as a limit)

A function $f(x)$ is *continuous* at $x = a$ if

$$\lim_{x \to a} f(x) = f(a).$$

"Really?" Tetra said. "That's it?"

"That's it." Miruka looked up at me. "So you finally made it."

Tetra looked up too. She hesitantly nodded my way.

"Let's see if this guy did a half decent job at teaching you about (ϵ, δ)," Miruka said. "Give this a shot."

Problem 6-1 (Defining continuity in terms of (ϵ, δ))

Show what it means for a function $f(x)$ to be continuous at $x = a$ in terms of (ϵ, δ).

"Hmm, let's see ..." Tetra said, tapping her chin. "Well, you said that a function being continuous at $x = a$ is defined like this."

$$\lim_{x \to a} f(x) = f(a)$$

"And we can also write that like this."

$$f(x) \to f(a) \quad \text{as} \quad x \to a$$

"In other words, as x gets arbitrarily close to a, $f(x)$ gets arbitrarily close to $f(a)$. I'm good so far?"

Tetra looked up at Miruka, who nodded and told her to go on.

"So I guess what I need to do is use (ϵ, δ) to rewrite this limit somehow. I may be biting off more than I can chew, but maybe if I just take it one piece at a time... Hang on while I give it a shot."

Tetra scribbled silently in her notebook, then paused and ran a finger along what she'd written.

"Okay, I think I've got it. I just need to plug $f(a)$ into the limit part of the (ϵ, δ) definition, right?"

$$\forall \epsilon > 0 \; \exists \delta > 0 \; \forall x \; \left[0 < |x - a| < \delta \Rightarrow |f(x) - f(a)| < \epsilon \right]$$

"In other words," she continued, "for any positive number ϵ we can choose some positive number δ so that for any value of x, the statement $0 < |x - a| < \delta \Rightarrow |f(x) - f(a)| < \epsilon$ is true."

"Good," Miruka said. "Which means?"

"That for any ϵ, we can choose a δ so that if x is in the δ-neighborhood of a, then $f(x)$ is in the ϵ-neighborhood of $f(a)$."

"I've got to admit, I'm impressed," Miruka said.

"Whew! I'm glad I practiced so much last night!" Tetra said, glancing my way.

Answer 6-1 (Defining continuity in terms of (ϵ, δ))

A function $f(x)$ is *continuous* at $x = a$ if

$$\forall \epsilon > 0 \; \exists \delta > 0 \; \forall x \; \Big[0 < |x-a| < \delta \Rightarrow |f(x)-f(a)| < \epsilon \Big]$$

"This is another case where I prefer graphs, though," Tetra said. "They make everything so much clearer! I mean, if a function isn't continuous it's going to be broken somewhere, so you can just *see* it, right? Like, if the point at $x = a$ had jumped off the curve, it's clearly a discontinuity."

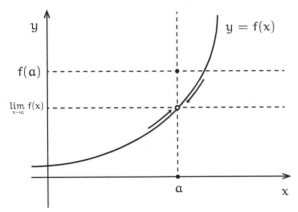

A function that isn't continuous at $x = a$

Miruka's mouth twisted into a mischievous smile.
"If only it were always that simple..."

6.4.2 *Continuous Nowhere*

"Answer me this," Miruka said.

Problem 6-2 (Discontinuity at all points)

Does there exist a function that is discontinuous for all real values?

"So what I'm looking for is a function that's continuous *nowhere*."

"Whoa," Tetra said. "I can't graph a function where every point is discontinuous."

"No, you can't. So don't try."

"But … but … I don't even know how to *imagine* a function like that!"

"You can't rely on imagination, either."

"So what should I rely on?"

"Logic."

"Logic?"

Miruka sighed.

"Tetra, why are you abandoning the definition so soon?"

"Oh! The (ϵ, δ) definition, right?"

"Of course. If you want to describe something that's *not* continuous at a, just negate the statement that says it *is* continuous there."

$$\neg \left(\forall \epsilon > 0 \; \exists \delta > 0 \; \forall x \; \left[0 < |a - x| < \delta \Rightarrow |f(x) - f(a)| < \epsilon \right] \right)$$

"In predicate logic, if you swap 'there exists' and 'for all' signs, you can force a 'not' into an expression. Like this."

$$\neg \left(\forall x \; \left[\cdots \right] \right) \quad \Longleftrightarrow \quad \exists x \; \left[\neg \left(\cdots \right) \right]$$
$$\neg \left(\exists x \; \left[\cdots \right] \right) \quad \Longleftrightarrow \quad \forall x \; \left[\neg \left(\cdots \right) \right]$$

"Do that to our (ϵ, δ) expression, and you get this."

$$\exists \epsilon > 0 \; \forall \delta > 0 \; \exists x \; \left[\neg \left(0 < |a - x| < \delta \Rightarrow |f(x) - f(a)| < \epsilon \right) \right]$$

"In other words, 'it's possible to choose some positive number ϵ such that no matter how small you make a positive number δ, there is some positive number x for which $0 < |x - a| < \delta \Rightarrow |f(x) - f(a)| < \epsilon$ does not hold.' This is the definition of $f(x)$ being *discontinuous* at $x = a$. So all you have to find is some function $f(x)$ where this is true for all real numbers a."

Tetra placed a hand on either side of her head and gave a soft groan.

"What do *you* think?" Miruka asked, turning to me.

Her voice, the sound of her talking math, was exactly what I needed to sweep away the last vestiges of my depression. I was back in the game.

"Here's a classic," I said.

$$f(x) = \begin{cases} 1 & \text{(when } x \text{ is an irrational number)} \\ 0 & \text{(when } x \text{ is a rational number)} \end{cases}$$

"Indeed."

"Hang on," Tetra said. "Is that really a function?"

"Sure," I said. "You plug in a number, you get a number back. That's what a function is. Think of this one as an irrational number detector."

Answer 6-2 (Discontinuity at all points)

There *does* exist a function that is discontinuous for any given real value.

6.4.3 Continuous at Only One Point

"No fair," Tetra said. "He already knew the answer."

"Yeah, I read about it somewhere," I admitted.

"Let's up the stakes, then," Miruka said, "and see how he does with this one."

Problem 6-3 (Continuity at one point)

Does there exist a function that is continuous only at $x = 0$?

"I dunno, that doesn't seem possible, somehow," I said, scratching my head.

"Um ..." Tetra tentatitvely raised a hand.

"Did you have a question?" Miruka asked.

"No, the answer, I think."

"Oh yeah? Let's hear it."

"Hold up!" I said. "Give me some time to think first!"

"Sure, no problem," Tetra said.

"Tetra, come tell me." Miruka cupped a hand to the side of her head.

"Uh, sure."

Tetra trotted over to where Miruka was sitting and bent down to whisper in her ear.

"Correct," Miruka said.

"Yay!" Tetra squealed, making my blood boil.

I went with my instincts, thinking how I could show that such a function could not possibly exist. There had to be a good reason out there, somewhere. I just didn't have it yet.

"I'll give you five more minutes," Miruka said.

6.4.4 Escape from an Infinite Maze

Miruka and Tetra continued a whispered conversation as I thought.

"Tell me how you studied this," Miruka said.

"Mainly I just kept rewriting the (ϵ, N) and (ϵ, δ) definitions, thinking about what they meant. Oh, and the graphs that show what a neighborhood is, too."

"Hmph."

"I'm pretty used to (ϵ, δ), now. There's still a few things I'm not totally getting, though."

"Your doubts may be coming from the nature of the real numbers themselves, which are tricky things to define with precision. You may be better off just staying at the 'not totally getting it' stage for now. Better that than falsely convincing yourself that you do get it."

"One question, though. I understand that with limits the guarantee that a δ exists is important, but even when I read the (ϵ, δ) definition I don't really get a feeling of 'getting arbitrarily close.' It still seems like getting 'arbitrarily close' should require infinitely many repetitions."

"That's not a good line of thinking—it leads you into a maze from which there's no escape. It makes it seem like you'll never attain your goal, no matter how far you go. But as much as I love mazes, there's no infinite repetition going on here, thanks to the guarantee that there's a δ out there for any ε that might come your way."

"Okay."

"Every person on earth who wants to study advanced mathematics is required to receive from Weierstrass this key called (ε, δ). It's the only way to open the door of limits and escape from this infinite maze."

6.4.5 *Finding the Function*

"Time's up," Miruka said.

"I give up," I said. "I still don't think any such function exists, but I can't figure out how to show that."

Miruka looked at Tetra, prompting her to talk.

"Actually ... I found one. So yeah, they do exist."

"You found one? No way."

"But all I did was tweak the function you came up with."

$$g(x) = \begin{cases} x & \text{(when } x \text{ is an irrational number)} \\ 0 & \text{(when } x \text{ is a rational number)} \end{cases}$$

"What? How ... ?"

"Explain," Miruka said.

"Sure," Tetra said. She turned to me. "You created that function $f(x)$, that isn't continuous anywhere, right? Your irrational number detector. Well, it's impossible to graph that, but I was kinda wondering what it would look like if you *could*. I think you'd get something like this."

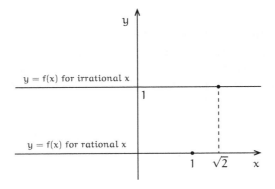

"It looks like two lines," Tetra continued, "but really it's the graph of a single function. Like, if $x = 1$ then $y = 0$, but for $x = \sqrt{2}$ the y-value is 1. Anyway, I wondered if I couldn't use this to somehow create a function like in Miruka's problem, where it's only continuous where $x = 0$. So I thought about the definition of continuity—this (ϵ, δ) one here. I wondered if I couldn't create a function $g(x)$ where no matter how small an ϵ shows up, we could still guarantee the existence of a δ that would put $g(x)$ in the ϵ-neighborhood of $g(0)$, so long as x was in the δ-neighborhood of 0."

Tetra glanced at me, and I nodded to indicate that I was following her.

"When x is rational $g(x) = 0$, so $g(x)$ is in the ϵ-neighborhood, no problem. The problem is the irrationals. So I started wondering how we could get $g(x)$ to be *really, really close* to $g(0) = 0$ when x is irrational. Then I realized, sure, we can do that if we tilt the previous graph so that it runs diagonal and gets closer that way."

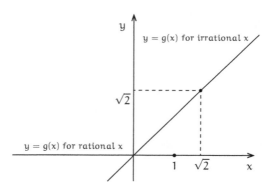

"When things are set up like this, then no matter how small the ϵ, we can definitely find a δ, since all we need to do is make the δ smaller than ϵ. Like, by using $\delta = \frac{\epsilon}{2}$, for example. If we do that, then any x in the δ-neighborhood will always put $g(x)$ in the ϵ-neighborhood."

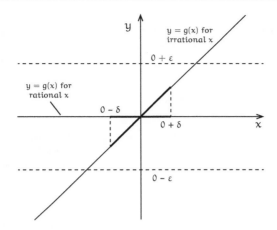

"Any rational x will just give 0, so we're okay there. But even if it's an irrational, then $\left|0 - g(x)\right| = \left|g(x)\right|$, which is smaller than ϵ. Like this, right?"

$$\left|g(x)\right| = \left|x\right| \qquad \text{when } x \text{ is irrational, } g(x) = x$$

$$< \delta \qquad \text{because } x \text{ is in the } \delta\text{-neighborhood}$$

$$= \frac{\epsilon}{2} \qquad \text{defined as } \delta = \frac{\epsilon}{2}$$

$$< \epsilon \qquad \epsilon > 0, \text{ so } \frac{\epsilon}{2} < \epsilon$$

"So in the end we get $\left|g(x)\right| < \epsilon$, which means $g(x)$ is in the ϵ-neighborhood of 0, which is exactly what we wanted! No matter how small the ϵ, if x is in the δ-neighborhood, then $g(x)$ is in the ϵ-neighborhood. And there it is—a function that's continuous only at $x = 0$."

"Too early to claim that," Miruka said. "You haven't shown that $g(x)$ is discontinuous when $x \neq 0$."

"Oops. I hadn't thought about that."

"Don't worry about it for now. It's easy enough to show."

"This was a cool problem. And you were right, Miruka—using words like 'arbitrarily close' doesn't quite cut it if you want to talk about whether this function $g(x)$ is continuous. I mean, you can't even graph it, not in a true sense at least. But still, imagining what the graph *might* look like was helpful. Thinking about that along with the (ϵ, δ) definition is how I came up with the function!"

Tetra sat back in her chair, smiling.

"Well done," Miruka said, patting her on the head.

Answer 6-3 (Continuity at one point)

A function that is continuous only at $x = 0$ *does* exist.

I started to say something, but somehow the words just wouldn't come out. Instead I mumbled some excuse and left the library.

6.4.6 Realizations

I headed back to my classroom to pick up some stuff I'd left behind, then cut through the courtyard on my way home. I didn't quite have the energy to make it that far, though, so I flumped onto a bench and put my head in my hands.

What is wrong *with me?*

Is it really such a big deal that I didn't ace a test? That Tetra showed me up at math? Is that really all it takes to turn me into a brooding idiot?

I heard footsteps behind me.

"You okay?"

Tetra.

I remained as I was, my face still in my hands.

"Are you sick or something?"

"Just disgusted with myself."

Tetra was silent for a time, then I felt her hand on my head. Her sweet smell. Her breath on my ear, as she leaned in closer and whispered.

"Thank you for showing me the joy of mathematics. I never would have experienced it without you. I hope you go on to help many, many more people feel it too."

The joy of mathematics.

It was a feeling I knew well. The triumph of solving a difficult problem. The fulfillment of perceiving new structures. The thrill of finding bridges between worlds. The delight of receiving messages sent by mathematicians centuries ago.

Yes, there is hardship in studying mathematics, but there's joy as well.

Just as there's joy in helping others to learn it.

That was when I realized something I'd known all along—that I was a teacher.

Normally I would have been mortified to let someone see me cry, but that day it didn't matter. I rubbed at my eyes, fixed my glasses, and looked up at Tetra.

"I'm sorry for the way I've been acting. Thanks for pulling me back."

Tetra smiled and shook her head.

"That's what friends are for."

This more precise definition of limits, and from that the ability to truly provide automatic judgment of the correctness of proofs in analysis, was made possible by Weierstrass's lectures on analysis at Berlin University, where he introduced the (ϵ, δ)-definition of limits.

GÖDEL, HAYASHI, YASUGI
The Annotated Incompleteness Theorems

CHAPTER 7

Diagonalization

Replacing a variable "Replacing a variable x through
self-reference is called diagonalization" through
self-reference is called diagonalization.
You cannot prove a statement that diagonalizes "You
cannot prove a statement that diagonalizes x."

The Fictitious Book of "The Fictitious Book of x*"*

7.1 SEQUENCES OF SEQUENCES

7.1.1 *Countable Sets*

"There you are!" Tetra said. "I've got something for you."

I grunted and fumbled to hit the button on my stopwatch.

"Oh, sorry. Were you in the middle of something?"

"Don't worry about it."

I exhaled long and slow, trying to get my head out of the math world and back into reality. When I'm deep into a math problem, I tend to forget where I am. Sometimes I even forget when I am and *who* I am. Petty things like that tend to lose meaning.

I took stock of my surroundings.

It's after school. I'm in the library. It's February, just a month until the end of the school year, and two months before Miruka and I become seniors. Wow, time is passing fast...

I shook my head. "It's fine. Just practicing working problems faster. You said you have something for me?"

"Special delivery, straight from Mr. Muraki!"

Tetra held up an index card with a flourish.

"Oh, cool. I wonder what he's thought up for us this time."

Problem 7-1

Show that the set $\mathbb{R}$ of all real numbers is not a countable set.

"Hey, I know this problem," I said. "It shows up in math books all the time."

"It's that famous?"

"Yeah, you use Cantor's diagonal argument. Want to hear about it?"

"Only if you're sure I'm not interrupting."

"Not at all."

Tetra sat in the chair to my left.

"Actually it's a pretty quick explanation," I said. "But before that, do you understand what the problem says?"

"Everything except 'countable set,' I guess."

"Easy enough. A countable set is just one where you can assign a natural number to each of its elements."

Definition of countable set

A set for which it is possible to assign a different natural number to each of its elements is called a *countable set*.

"Okay, next question," Tetra said. "What does it mean to be able to assign natural numbers?"

"Well, for example any finite set is countable.[1] Since there are

---

[1] Note that some authors prefer not to include finite sets in the definition of countable sets. Where distinguishing between the two is preferred, the phrase "countably infinite" refers to sets with the same cardinality as $\mathbb{N}$, and "at most countable" describes sets that are either finite or countably infinite.

only a finite number of elements, you can just label each one with a natural number."

"In other words, you can count them."

"In that case, yeah, but you can't 'count' an infinite set by the usual definition. That doesn't mean it's not a countable set, though."

"I'm getting confused..."

"Okay, let's look at the set of all integers $\mathbb{Z}$ as an example of an infinite set."

$$\mathbb{Z} = \{\,\ldots,\ -3,\ -2,\ -1,\ 0,\ +1,\ +2,\ +3,\ \ldots\,\}$$

"This is a countable set," I said, "because we can assign a unique natural number to each element. One way would be to assign the natural number 1 to the integer 0, then 2 to +1, 3 to −1, 4 to +2, and so on."

1	2	3	4	5	6	$\cdots$	$2k-1$	$2k$	$\cdots$	all natural numbers
↓	↓	↓	↓	↓	↓		↓	↓		
0	+1	−1	+2	−2	+3	$\cdots$	$1-k$	$+k$	$\cdots$	all integers

"Um, okay..."

"Maybe it's easier to see if I write it like this."

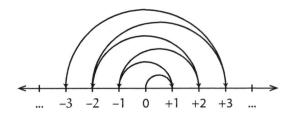

"Ah, so you're just bopping back and forth between the positives and negatives," Tetra said.

"Right. It doesn't really matter how you do it, though, so long as every integer gets labeled with its own, unique natural number. Since we've found a way to do that, we know that $\mathbb{Z}$ is a countable set."

"Okay, I think I got that. And I assume there are other infinite-but-countable sets out there?"

"Sure. Take the set of all rational numbers $\mathbb{Q}$, for example. You can trace through every rational number 0 and up by doing something like this."

$$
\begin{array}{ccccccccc}
0 & \to & \frac{1}{1} & \to & \frac{2}{1} & & \frac{3}{1} & \to & \frac{4}{1} & & \cdots \\
& & & \swarrow & & \nearrow & & \swarrow & & \nearrow \\
& & \frac{1}{2} & & \frac{2}{2} & & \frac{3}{2} & & \frac{4}{2} \\
& & \downarrow & \nearrow & & \swarrow & & \nearrow \\
& & \frac{1}{3} & & \frac{2}{3} & & \frac{3}{3} \\
& & & \swarrow & & \nearrow \\
& & \frac{1}{4} & & \frac{2}{4} \\
& & \downarrow & \nearrow \\
& & \frac{1}{5}
\end{array}
$$

"Then you can include the negatives by just alternating back and forth, like we did with the integers."

$$0 \to +\frac{1}{1} \to -\frac{1}{1} \to +\frac{2}{1} \to -\frac{2}{1} \to +\frac{1}{2} \to -\frac{1}{2} \to +\frac{1}{3} \to -\frac{1}{3} \to \cdots$$

"If you really wanted to be picky you could skip duplicates that pop up after reduction, like $\frac{2}{2}$ after $\frac{1}{1}$, but it's the same idea."

"So this means that the rational numbers are countable." Tetra nodded, then cocked her head. "But isn't that how it would work for *any* set? After all, there's infinitely many integers to hand out, so how could you end up not having enough?"

"If that were the case, Mr. Muraki would be handing us a bad problem. Read it again—he wants us to show that the set of all real numbers is *not* countable. In other words, there's no way you can assign a different natural number to every real number, despite there being infinitely many natural numbers you can use."

"That seems...weird. Why can't you just take the next real, label it with the next natural number, and keep going on and on?"

"Because you aren't guaranteeing that a number would be assigned to *every* real number. For one thing, how would you decide what the 'next real number' is?"

"I would...uh, I don't know. But just because *I* don't know doesn't mean there's no way to do it. Maybe somebody else has, or will, think of a good way. How can you say for certain that it's impossible?"

"The same way we say anything with certainty in mathematics—using a proof."

7.1.2 Cantor's Diagonal Argument

Problem 7-1
Show that the set $\mathbb{R}$ of all real numbers is not a countable set.

"Like I said, we can give a proof using something called Cantor's diagonal argument, but to do that we have to change the form of the problem a little bit."

"Isn't that cheating?" asked Tetra.

"Not in this case. Instead of attempting to assign natural numbers to *all* the real numbers, let's think about just assigning them to real numbers between 0 and 1. It's basically the same thing."

"How?"

"Well, take a look at this graph."

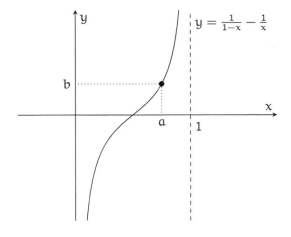

"What does this show?" she asked.

"That we can choose any real value a in the range $0 < x < 1$ on the x-axis and get a corresponding real value b on the y-axis. Or we can go the other way, starting from a real value b on the y-axis and getting a real number a in the range $0 < x < 1$ on the x-axis. So there's a one-to-one correspondence—a bijection, remember?—between the values $0 < x < 1$ and the real numbers. It works exactly the same way as if we were going to label all the real numbers."

Problem 7-1a (Problem 7-1 rephrased)
Show that the set of all real numbers $0 < x < 1$ is not a countable set.

"So let's talk about Cantor's diagonal argument. It uses proof by contradiction. Remember that?"

Tetra nodded. "The one where you assume that what you want to prove is false, then figure out how that results in a contradiction."

"Exactly. The proposition we want to prove is 'the set of all real numbers $0 < x < 1$ is *not* a countable set,' so we assume its negation."

> The set of all real numbers $0 < x < 1$ *is* a countable set

"We want to start from that—in other words from the assumption that we can assign a unique natural number to each of the real numbers in that range, and show how that leads to a contradiction."

"Got it!"

"Okay then, let's assume that we have such a labeling scheme, and call the labels A_n. For example, maybe our A_n's look something like this."

$$\begin{cases} A_1 &= 0.01010\cdots \\ A_2 &= 0.33333\cdots \\ A_3 &= 0.14142\cdots \\ A_4 &= 0.10000\cdots \\ A_5 &= 0.31415\cdots \\ \vdots & \quad \vdots \end{cases}$$

"We can generalize writing real numbers 'zero-point-something' like this."

$$A_n = 0 \,.\, a_{n,1}\ a_{n,2}\ a_{n,3}\ a_{n,4}\ a_{n,5}\ \cdots$$

"What's with the double subscripts?"

"We want to know where in the nth number each digit is appearing. So $a_{n,1}$ is the first digit after the nth number's decimal, $a_{n,2}$ is the second number, and so on. So to generalize, $a_{n,k}$ is the kth digit after the decimal in the real number A_n. Taking $A_5 = 0.31415\cdots$ as an example, we get this."

$$a_{5,1} = 3,\ \ a_{5,2} = 1,\ \ a_{5,3} = 4,\ \ a_{5,4} = 1,\ \ a_{5,5} = 5,\ \ \ldots$$

Tetra traced a finger along the series of numbers.

"Maybe it would be easier to follow if I broke it down like this," I said.

$$\begin{array}{ccccccccc} A_5 &=& 0 &.& 3 & 1 & 4 & 1 & 5 & \cdots \\ & & & & \shortparallel & \shortparallel & \shortparallel & \shortparallel & \shortparallel & \cdots \\ & & & & a_{5,1} & a_{5,2} & a_{5,3} & a_{5,4} & a_{5,5} & \cdots \end{array}$$

"Much better!" Tetra said. "Okay, I'm caught up."

"Great. I just remembered another condition, though. Some reals can be written in two different ways. For example, $0.1999\cdots$ and $0.2000\cdots$ actually represent the same number. Let's avoid any confusion by saying that we won't use real numbers ending with an infinitely long string of nines."

"Fair enough."

"Also, don't forget that we're working in the range $0 < x < 1$, so we won't be looking at $0.000\cdots$, either."

"Okay, but I'm still not even sure what we're looking at to begin with."

"I was just getting to that. Hang on, let me create a table of A_n values with their digits labeled ..."

		1	2	3	4	5	...
$A_1 =$	0.	$a_{1,1}$	$a_{1,2}$	$a_{1,3}$	$a_{1,4}$	$a_{1,5}$	...
$A_2 =$	0.	$a_{2,1}$	$a_{2,2}$	$a_{2,3}$	$a_{2,4}$	$a_{2,5}$	...
$A_3 =$	0.	$a_{3,1}$	$a_{3,2}$	$a_{3,3}$	$a_{3,4}$	$a_{3,5}$	...
$A_4 =$	0.	$a_{4,1}$	$a_{4,2}$	$a_{4,3}$	$a_{4,4}$	$a_{4,5}$	...
$A_5 =$	0.	$a_{5,1}$	$a_{5,2}$	$a_{5,3}$	$a_{5,4}$	$a_{5,5}$	...
$\vdots$		$\vdots$	$\vdots$	$\vdots$	$\vdots$	$\vdots$	$\ddots$

"So say this is a list of all our A_n values," I said. "We're working on the assumption that the set of all real numbers between 0 and 1 is a countable set, so if we've listed all of the A_n's, this should be a list of *every real number* $0 < x < 1$, right?"

"Sure."

"Okay, so let's look at the numbers running down the diagonal of this table."

		1	2	3	4	5	...
$A_1 =$	0.	$\underline{a_{1,1}}$	$a_{1,2}$	$a_{1,3}$	$a_{1,4}$	$a_{1,5}$	...
$A_2 =$	0.	$a_{2,1}$	$\underline{a_{2,2}}$	$a_{2,3}$	$a_{2,4}$	$a_{2,5}$	...
$A_3 =$	0.	$a_{3,1}$	$a_{3,2}$	$\underline{a_{3,3}}$	$a_{3,4}$	$a_{3,5}$	...
$A_4 =$	0.	$a_{4,1}$	$a_{4,2}$	$a_{4,3}$	$\underline{a_{4,4}}$	$a_{4,5}$	...
$A_5 =$	0.	$a_{5,1}$	$a_{5,2}$	$a_{5,3}$	$a_{5,4}$	$\underline{a_{5,5}}$	...
$\vdots$		$\vdots$	$\vdots$	$\vdots$	$\vdots$	$\vdots$	$\ddots$

"If we create a sequence out of those numbers on the diagonal, we get this."

$$a_{1,1}, \ a_{2,2}, \ a_{3,3}, \ a_{4,4}, \ a_{5,5}, \ \ldots$$

"Next, we use this sequence $\langle a_{n,n} \rangle$ to create a new sequence $\langle b_n \rangle$, like this."

$$b_n = \begin{cases} 1 & \text{when } a_{n,n} \text{ is one of } 0, 2, 4, 6, 8 \\ 2 & \text{when } a_{n,n} \text{ is one of } 1, 3, 5, 7, 9 \end{cases}$$

"In other words if $a_{n,n}$ is even, then b_n is 1, and if $a_{n,n}$ is odd, then b_n is 2. Do you see how that would make this true for all natural numbers n?"

$$b_n \neq a_{n,n}$$

Tetra thought for a moment, then said, "Because you're assigning an odd number where there were even digits, and even ones for odd digits. So b_n can't be $a_{n,n}$, because the parity of every digit is flipped."

"Perfect. Okay then, let's define a real number B like this."

$$B = 0 . b_1 \, b_2 \, b_3 \, b_4 \cdots$$

Tetra grimaced. "*Another* new definition?"

"Let's work with a specific example," I said. "First we go back to the labels we said we'd assigned, and pull out the values on its diagonal."

		1	2	3	4	5	...
$A_1 =$	0.	$\underline{0}$	1	0	1	0	...
$A_2 =$	0.	3	$\underline{3}$	3	3	3	...
$A_3 =$	0.	1	4	$\underline{1}$	4	2	...
$A_4 =$	0.	1	0	0	$\underline{0}$	0	...
$A_5 =$	0.	3	1	4	1	$\underline{5}$	...
$\vdots$		$\vdots$	$\vdots$	$\vdots$	$\vdots$	$\vdots$	$\ddots$

"Then sequence $\langle a_{n,n} \rangle$ looks like this."

$$0,\ 3,\ 1,\ 0,\ 5,\ \ldots$$

"Next we create the sequence $\langle b_n \rangle$ by replacing even numbers with 1, and odd numbers with 2."

$$1,\ 2,\ 2,\ 1,\ 2,\ \ldots$$

"Now we have our real number B."

$$B = 0.12212 \cdots$$

"Do you see how B will always be in the range $0 < x < 1$, no matter what list of A_n values we started with?"

"Sure," Tetra said. "Because it's always going to be zero-point-a-bunch-of-ones-and-twos."

"Right. And what does that say about B, regarding our list of labels?"

"That it has to be in there, somewhere?"

"Exactly! And that's very important. Since it has to be there somewhere, let's say it's in row m, which lets us write this."

$$A_m = B$$

"Now, let's zoom in on that part of our list."

			1	2	3	...	m	...
	$A_1 =$	0.	$\underline{a_{1,1}}$	$a_{1,2}$	$a_{1,3}$	...	$a_{1,m}$	...
	$A_2 =$	0.	$a_{2,1}$	$\underline{a_{2,2}}$	$a_{2,3}$	...	$a_{2,m}$	...
	$A_3 =$	0.	$a_{3,1}$	$a_{3,2}$	$\underline{a_{3,3}}$	...	$a_{3,m}$	...
	$\vdots$		$\vdots$	$\vdots$	$\vdots$	$\ddots$	$\vdots$	...
$B =$	$A_m =$	0.	$a_{m,1}$	$a_{m,2}$	$a_{m,3}$	...	$\underline{a_{m,m}}$	...
			$\|\|$	$\|\|$	$\|\|$	...	$\|\|$	...
			b_1	b_2	b_3	...	b_m	
	$\vdots$		$\vdots$	$\vdots$	$\vdots$	$\vdots$	$\vdots$	$\ddots$

"In particular, we want to take a look at the intersection of row m and the diagonal."

$$\underbrace{a_{m,m}}_{\text{the } m\text{th digit of } A_m} = \underbrace{b_m}_{\text{the } m\text{th digit of } B}$$

"Here we're looking at the digit in position m after the decimal in number A_m, which is also B. But think about how we created B in the first place, and how that implied that $a_{n,n} \neq b_n$."

"Sure." Tetra nodded. "After all, the whole point was setting things up so that b_n behaved that way."

"Indeed it was. But if we're going to say that $a_{n,n} \neq b_n$ holds for all natural numbers n, it has to work for natural number m too. So we get this."

$$a_{m,m} \neq b_m$$

"And this is...?"

"A contradiction!" Tetra nearly shouted, her eyes wide.

"Which is exactly what we wanted. We can now say that proof by contradiction shows that the set of real numbers $0 < x < 1$ is *not* a countable set."

Answer 7-1a

Use proof by contradiction.

1. Suppose that the set $S = \{x \in \mathbb{R} \mid 0 < x < 1\}$ is a countable set.

2. We can represent an arbitrary element of set S as

$$A_n = 0 . a_{n,1} \, a_{n,2} \, a_{n,3} \, a_{n,4} \, \cdots \, a_{n,k} \, \cdots .$$

3. Define real number B as

$$B = 0 . b_1 \, b_2 \, b_3 \, b_4 \, \cdots \, b_n \, \cdots ,$$

where b_n is defined as

$$b_n = \begin{cases} 1 & \text{when } a_{n,n} \text{ is even,} \\ 2 & \text{when } a_{n,n} \text{ is odd.} \end{cases}$$

4. From the definition of b_n, $a_{n,n} \neq b_n$ for any natural number n.

5. B is an element of set S, so there exists an m such that $A_m = B$.

6. Then for the mth digit of B, we have $a_{m,m} = b_m$.

7. But from 4. above, we have that $a_{m,m} \neq b_m$.

8. Items 6. and 7. are a contradiction.

9. By proof by contradiction, we thus have that set S is not a countable set. □

"Very cool," Tetra said. "So this is the answer to Mr. Muraki's card?"

"Well, not quite. We changed things around, remember? But it lets us give an answer. We can create a one-to-one correspondence between the set of real numbers $0 < x < 1$ and the set $\mathbb{R}$ of all real numbers, like in the graph I drew when we started. So now that we've shown that the set of real numbers $0 < x < 1$ is not a countable set, we can say that $\mathbb{R}$ isn't one, either."

Answer 7-1

The set of real numbers $0 < x < 1$ and the set of all real numbers $\mathbb{R}$ are in one-to-one correspondence. Therefore, from Answer 7-1a, $\mathbb{R}$ is not a countable set.

"I think that does it," I said.

Tetra sat thinking silently for a while, then raised her hand.

"Question! This is called Cantor's diagonal argument because we use diagonal entries in the table, right?"

"Right. It's infinitely large, so you never run out of rows or columns, but I guess we can still call it a table."

"I think I get the gist of what's going on, but I still have a question."

"Fire away."

"If the real number B isn't in the table, can't we just add it?"

"That wouldn't help, because the contradiction appears as soon as we find B missing. Say you did add B, creating a new 'fixed' table. You could still just start over again, and by the same reasoning find another number C that wasn't there."

"Oh, right! And if you added C you could still find a missing D, and so on."

"That's right."

"How is it that you always know the answers to my questions as soon as I ask them?"

"Because I had the same questions myself, once. But I've studied this stuff a good bit, so I know it pretty well now."

"Hmph. I wonder . . ." came a familiar voice from behind us.

Tetra yelped and I spun around.

Sure enough, it was Miruka.

7.1.3 A Challenge

"*Please* try to make some noise when you walk up on us," I said. "You're going to give someone a heart attack."

"Statistically speaking that's very unlikely," Miruka said. "More to the point, did you just say you know diagonalization well?"

"Uh...I did."

With Miruka standing there, hand on hip and staring at me, saying so felt like confessing to a crime.

Miruka shook her head. "Muraki strikes again."

"What are you talking about?" I said.

"He gave me some very specific instructions—he said, 'After he teaches Tetra about diagonalization, he'll say something about how well he knows it. When he does, hand him this card.'"

Miruka put a card on the table in front of me and sat down on my right.

Problem 7-2 (A challenge: Enumerating reals)

Demonstrate the truth or falsity of the following proof:

We can enumerate real numbers beginning with '0.' as follows:

There are only ten possibilities for the digit in the first place after the decimal. It is therefore possible to create a complete labeling of sequences up to the first digit after the decimal. For each of those ten possibilities, there are only ten possibilities for the subsequent digit. It is therefore possible to create a complete labeling of sequences up to the second digit after the decimal. We can repeat this process of creating complete listings, regardless of how many digits follow the decimal. Thus, the set of all numeric sequences beginning with '0.' is a countable set.

"I have no idea what this is all about," Tetra said, peering at the card.

"If we can label every element in a set with a different natural number, it's a countable set," Miruka said. "If this method works, then the set of all real numbers between 0 and 1 is a countable set. But you've just shown that it *isn't* a countable set."

"Well then this new problem is easy—it doesn't work!"

"I'd say you're probably right. But the question is, *why* doesn't it work?"

"And that's not necessarily a simple question," I said. "I mean, it kind of makes sense, right? Instead of just thinking about a random jumble of real numbers, maybe organizing things by starting smaller—with finite-length reals that we know we can exhaustively label—and working your way up from there is a valid strategy."

Miruka raised an eyebrow.

"No, I don't think it really works," I hurried to add, "but off the top of my head I can't say why."

I pulled out some scratch paper and started making lists of increasingly long real numbers while I thought.

- 0.0, 0.1, 0.2, ..., 0.9, (10 cases)

- 0.00, 0.01, 0.02, ..., 0.99, (100 cases)

- 0.000, 0.001, 0.002, ..., 0.999, (1000 cases)

- 0.0000, 0.0001, 0.0002, ..., 0.9999, (10000 cases)

- Etc., etc.

"Well, the problem's definitely right about there being ten possibilities for each digit, and I don't see how any could be missed when labeling them, no matter how many you add..."

"Maybe there's something wrong with your diagonalization," Miruka said. "Maybe $0 < x < 1$ does give a countable set after all."

Her delivery didn't give it away, but her eyes made it clear she found the situation hilarious. Math humor, Miruka-style.

Tetra raised her hand.

"Miruka, can I ask a question?"

"That's a meta-question."

"Oh, a question about questions! You're right!" Tetra grinned.

"Never mind," Miruka said. "What did you want to ask?"

"About this method... It seems like it would assign a label to 0, which is outside the range $0 < x < 1$. There's also all kinds of duplicated numbers, like 0.01 and 0.010 and 0.0100. They're equal numbers, but they're being counted more than once. Is that why this method doesn't really work?"

"Those are good things to notice, but they aren't really a problem here. You can just skip anything that's out of range or that's already been counted. Didn't you do that with the rationals?"

"Now that you mention it..."

I was starting to panic. This felt like a problem that would have an immediate answer, but I wasn't seeing it. Diagonalization was a classic topic in lots of math books, and it was something that I had *thought* I understood well.

So why can't I find the error?

Tetra was clearly thinking hard too, which turned up the pressure. Even so, my approach of considering what happened with more and more digits wasn't taking me anywhere...

Wait, is that the key to this whole thing?

Sure, you could use this method to exhaustively label real numbers, *so long as they had finite length.*

But that's not the only kind of real number. For instance, 0.333... is a perfectly fine real number but one with infinitely many digits. The method described on Mr. Muraki's card would never label it!

"I got it," I said. "The method fails because it can only label real numbers with a finite number of digits."

"Yep," Miruka said.

"Hey, no spoilers!" Tetra shouted.

"Oops, sorry," I said. "But that's pretty much it. Some real numbers are infinitely long. Plenty of rationals too. That's even the case in the range $0 < x < 1$. Like one-third, for example."

$$\frac{1}{3} = 0.333\ldots$$

"Or pi over ten."

$$\frac{\pi}{10} = 0.314159265\ldots$$

"Sure, Mr. Muraki's proposed method works for arbitrarily long numbers, but only for ones with finite length. It doesn't apply to numbers like this, so it doesn't cover the entire set."

Answer 7-2 (A challenge: Enumerating reals)
The argument is **false**, because it does not enumerate real numbers with infinitely many nonzero digits.

Miruka rewarded me with a nod. I returned a smile, happy to have beaten Mr. Muraki's second challenge.

"By the way," Miruka said, "Muraki did say one other thing..."

"What's that?" I asked.

"Something along the lines of, 'He'll probably solve this one pretty quick and get all puffed up about it. When he does, tell him to flip the card over.'"

7.1.4 Another Challenge

I groaned, staring at the card in front of me with dread.

No use putting it off...

I flipped the card over, and sure enough, there was another problem.

Problem 7-3 (A challenge: Diagonalizing the rationals)
Find the error in this statement:
Replacing all occurences of the word "real" in Cantor's diagonal argument with the word "rational" shows that the set of all rational numbers is not a countable set.

"Hmm..."

"What's all this about?" Tetra asked.

"Just what it says," Miruka said. "Why can't you apply this argument to the rationals? Let A_n represent rational numbers instead of reals, and assume you have a list of all of them between 0 and 1. Create a rational number B out of the numbers on the diagonal, one that doesn't exist in the list, and there you have it—the rationals aren't countable."

"But they are," I said.

"Show me."

There was still amusment in Miruka's eyes. I tore my eyes away from her so that I could concentrate on the problem. The rationals were in danger of being classified as not countable; they needed my help.

"Whatcha think, Tetra?" Miruka asked.

"I . . . have no idea," she said, shaking her head. "I guess there's got to be some difference between the reals and the rationals that make the argument work for one, but not the other."

"An excellent observation," Miruka said.

"Hey, yeah," I said. "That's what this problem must be all about—the difference between reals and rationals. We just have to figure out what that difference is, and how it applies to this proof."

So what is the difference? All rationals are reals too, just ones that can be represented as a fraction. Though I guess we're representing them in decimal in this case. To convert them to decimal notation we have to—oh, hey!

"Got it!" I said.

"Do you now?" Miruka said.

"Yeah. The problem is at the end, where we pull out the diagonal $a_{n,n}$ digits to create a new number B—we can't guarantee that B will be rational. When you create a decimal out of a fraction, you always end up with some kind of looping pattern, a repeating decimal. Like how $\frac{1}{3}$ becomes $0.333\ldots$, repeated threes, or $\frac{1}{7}$ becomes $0.142857142857142857\ldots$, which is just 142857 repeated over and over. But when we create B from the diagonal there's no guarantee that you're going to get something like that. And that means you can't just replace 'reals' with 'rationals' in Cantor's method."

Miruka nodded.

"That'll do," she said.

Answer 7-3 (A challenge: Diagonalizing the rationals)

The constructed number B is not guaranteed to be a rational number, so diagonalization cannot be applied.

"I think I also get what Mr. Muraki was really aiming at with these problems," I said.

"What's that?" Tetra asked.

"The difference between knowing *about* something and really *knowing* something. That even for something famous like diagonalization, just having heard about it or read about it in a book doesn't mean you really know what's going on."

"I'm living proof of that!" Tetra said. "Speaking of which, it was great to see another proof by contradiction. I'm finally getting comfortable with that."

"Good. It's a powerful technique."

"The word 'contradiction' used to feel strange to me—I would always imagine some kind of catastrophic chaos. But now I see that in mathematics, it's just another word, one that describes a logic step."

"You could say the same about 'negation,'" Miruka said.

"Sure!" Tetra said. "As an everyday word 'negation' sounds so...well, *negative*. But that's not the case in math, is it. I don't feel like I've done something bad when I negate a logic statement."

"Yet another case where you can't get too hung up on the dictionary meaning of words when you're doing mathematics," I said.

"Oh, and another thing," Tetra said, "I'm learning to really appreciate everything all these mathematicians have left for us. Peano's axioms, Dedekind's definition of infinity, Weierstrass's (ϵ, δ), Cantor's diagonalization... They all lead to such amazing, beautiful things. Like chasing after Snow White's glass slipper!"

"I think that was...never mind."

7.2 SYSTEMS OF SYSTEMS

7.2.1 *Completeness and Consistency*

We fell into silence for a while, contemplating Mr. Muraki's card. I glanced at Miruka. She had formed a birdcage with her hands, staring at it intensely. She too had a pianist's fingers, but they were different from Ay-Ay's. Longer, finer, more delicate.

"I want to talk about formal systems some more," she said, out of the blue.

"More mechanical thinking?" Tetra asked.

"Not quite. The other day we were talking about formal systems of propositional logic. Today's topic is formal systems of arithmetic."

"There's more than one kind of formal system?"

"If you're using the definition we talked about, there are infinitely many."

"Oh."

Miruka narrowed her eyes.

"Based on that question, I'm going to assume you've forgotten what we talked about the other day."

Tetra reddened.

"Uh, yeah... Maybe."

"Hmph. All right, a quick review then."

Miruka paused, as if sorting out a deck of mental note cards.

"In a formal system you define something called a propositional formula, which is simply a finite sequence of symbols. We don't care what they mean. We also choose some number of propositional formulas to use as axioms, and we set up some inference rules, which is a way of turning propositional formulas into other propositional formulas."

Miruka grabbed my notebook and pencil and turned to a fresh page.

· Propositional formulas

· Axioms and inference rules

"So we can start with the axioms and initial propositional formulas, and use the inference rules to create some finite sequence of propositional formulas. That's called a proof, and the propositional formula at the end of a proof is called a theorem."

· Proofs and theorems

"Is it coming back to you now, Tetra?" Miruka asked.

"It is! Proofs in formal systems aren't like the ones we do in normal math, they're just formal proofs, finite sequences of propositional formulas. Like, that proof of $(A) \to (A)$, it was a sequence of five propositional formulas. I'm starting to remember all this now."

"Good. So depending on what sequence of symbols you use as propositional formulas, and depending on which propositional formulas you use as your axioms, and depending on what kind of inference rules you set up, you can create all kinds of formal systems. The formal system of propositional logic we talked about the other day is just one simple example. It's a fun thing to play with, but it lacks expressiveness."

"Expressiveness?" I asked.

"Sure. Like, it allows us to write simple statements like this ..."

$$(A) \to (A)$$

" ... but not something like this."

$$\forall m \, \forall n \, \left[(m < 17 \land n < 17) \to m \times n \neq 17 \right]$$

Tetra and I both stared at what Miruka had written for a while.

"Oh, I get it," I said. "This says that 17 is a prime. See? It's saying that you can't multiply two numbers together to get 17."

"Why the inequalities?" asked Tetra.

"Because without them, you could use 1×17."

"Oh, of course! That's an important part of the definition of prime numbers!"

"There you two go with all the meaning again ..." Miruka said.

"Ah, right," I said, wincing. "We aren't supposed to think about what we're doing."

"Oh, c'mon!" Tetra said. "You must've wrote this particular propositional formula to say that 17 is prime, even if we aren't supposed to think about that when we're using it in a formal system. I mean, if you interpret this correctly, what else is there?"

"A very interesting point," Miruka said. "We need to pay careful attention to that idea of 'correct interpretation.' There are fields in mathematics that consider interpretation, like formal semantics and model theory. By defining interpretation, you *can* apply meaning to formal systems. But there's no single 'correct' interpretation—you can apply different interpretations to any formal system, and its meaning will change according to the interpretation it's saddled with. Still, there are standard interpretations that get more commonly used."

"Like what?" Tetra asked.

"Like just now, when you two assumed that m and n had to be natural numbers. Or like when you assumed that the $\times$ operator indicated the product of two natural numbers. Or like when you interpreted the $\neq$ as meaning 'not equal.' Sure, this statement says that 17 is prime, *if you make all of those assumptions*, but what if m and n were reals? Given that interpretation, that wouldn't be what it says. So if you're going to apply meaning to a formal system, you've got to be sure to define what your interpretations will be."

"Huh," Tetra and I said in unison.

"Of course, Tetra's right. I did write that with the intent of saying that 17 is prime."

Miruka smiled and winked at Tetra.

"But anyway, back to the matter at hand. Like I said, a formal system of propositional logic doesn't allow statements like this."

$$\forall m \, \forall n \, \Big[(m < 17 \wedge n < 17) \rightarrow m \times n \neq 17 \Big]$$

"That's because formal systems of propositional logic lack a few necessary tools."

Miruka wrote a list in the notebook.

· There is no $\forall$ symbol.

· There is no symbol like $\times$ for expressing calculations between natural numbers.

· There are no symbols like $<$ and $\neq$ to show relations between numbers.

· There is no concept of using letters like m and n to indicate variables.

· There is no way of expressing natural numbers as constants, like 17.

"To formally express simple arithmetic operations like addition and subtraction of natural numbers, we'll need to add all of that."

"How do we do that?" Tetra asked.

"By introducing symbols, constants, and variables, and defining the required axioms and inference rules."

Tetra looked decidedly uneasy at the idea.

"That seems kinda ... chaotic. If anyone is allowed to just build up mathematics any way they like, it seems like there would be all kinds of weird math all over the place."

"No," Miruka said, emphatically shaking her head. "Anyone can create whatever formal systems they like, but it's not a problem. Just like lots of people creating bad music doesn't mean there won't be good music too. For one thing, there are important properties that a formal system needs to fulfill."

"Like what?"

"How about consistency? A 'good' formal system shouldn't contradict itself."

"How could it?"

"By allowing for some propositional formula A where you could prove both A and ¬A. If the system lets you say both, then it isn't consistent."

Definition of inconsistent formal systems

A formal system that allows proof of both A and ¬A for some propositional formula A (in other words, a system that contains contradictions) is *inconsistent*.

"I assume this means you could start from the axioms, and apply inference rules to reach both A and ¬A?" I asked.

"That's a good way to look at it," Miruka said. "Or you could say that if A and ¬A are among the many propositional formulas contained in the formal system, then it's inconsistent. You can also go glass-half-full and say that if it contains no such A, then it's consistent."

Definition of consistent formal systems

A formal system where there exists no propositional formula A for which both A and ¬A can be proved (in other words, a system that contains no contradictions) is *consistent*.

"Interesting," I said. "So formal systems let you talk about contradictions without resorting to concepts of 'true' and 'false.' The only thing that's important is what you can prove."

Tetra broke her long silence.

"But you have to be able to prove one or the other right? Either A or ¬A?"

"No," Miruka replied.

"No? Why not?"

"Because that's not what the definitions say," I said. "Look, compare these two statements."

- A and ¬A cannot both be proven.

- Only one of A or ¬A can be proven.

"Those are different?" Tetra asked.

"You're forgetting the third possibility," Miruka said.

"Huh? Oh, that neither one can be proved?"

"Right. Consistency doesn't necessarily require the proof of either A or ¬A. Just that they can't *both* be proved."

"Okay, I think."

"However," Miruka continued, "there's no limit to the number of propositional formulas containing free variables that can't be proved. But we're not interested in all possible propositional formulas, just the provability of ones that don't include free variables. Statements, in other words."

"What's a free variable?" Tetra asked.

"One that isn't bound by a symbol like $\forall$ or $\exists$. For example, the x that shows up three times in PF 1 here is a free variable. Since PF 1 contains free variables, it isn't a statement."

$$\forall m \, \forall n \, \Big[(m < x \wedge n < x) \to m \times n \neq x \Big] \quad \text{(PF 1: Not a statement)}$$

"On the other hand, PF 2 doesn't contain any free variables, so it is a statement."

$$\forall m \, \forall n \, \Big[(m < 17 \wedge n < 17) \to m \times n \neq 17 \Big] \quad \text{(PF 2: Is a statement)}$$

"Oh. Okay."

"So in a formal system of arithmetic," I said, "PF 1 is a *predicate* that says x is prime, and PF 2 is a *proposition* that says 17 is prime?"

Miruka nodded.

"But mainly I want you to remember that a statement is a propositional formula that contains no free variables."

"Got it," I said, and Tetra chimed in that she did too.

"So back to our formal system. Say there's some statement A for which you *cannot* prove either A or $\neg$A. In that case, A is called an undecidable statement, and a formal system that contains an undecidable statement is said to be incomplete. And of course, a formal system that isn't incomplete is called complete."

My eyes widened.

"Incomplete, as in—"

"As in Gödel's incompleteness theorems."

Incomplete formal systems

In a given formal system, if neither A nor $\neg$A can be proven for some statement A in the system, that system is *incomplete*.

Complete formal systems

In a given formal system, if at least one of A or ¬A can be proven for every statement A in the system, that system is *complete*.

"So say you've got a formal system," Miruka said. "Then if exactly one of A or ¬A can be proven for every statement A in the system, then that system is both consistent and complete, which is a wonderful thing to be. It's just the kind of thing that Hilbert was after, in fact."

Miruka paused, and took a deep breath.

"Too bad Gödel's incompleteness theorems shattered that dream."

7.2.2 Gödel's Incompleteness Theorems

"What are Gödel's incompleteness theorems?" asked Tetra.

"Two theorems about formal systems," Miruka said. "The first one says, in short, something like this."

Gödel's first incompleteness theorem

All formal systems meeting certain criteria are incomplete.

"We can rewrite it using our definition of incompleteness like this."

Gödel's first incompleteness theorem (rephrased)

All formal systems meeting certain criteria contain some statement A for which the following hold:

· A cannot be proven.

· ¬A cannot be proven.

"So you can't prove A," Tetra began, "but you can't prove its negation, either."

"Well, I'm using sloppy language here when I say 'cannot be proven.' What I mean is that you can't provide a formal proof. That of course raises an obvious question: what's the difference between a 'proof' and a 'formal proof'? So let me head that one off by rewriting this one more time."

Gödel's first incompleteness theorem (re-rephrased)

Formal systems meeting certain criteria contain some statement A for which the following hold:

- There exists no formal proof of A within that formal system.

- There exists no formal proof of $\neg A$ within that formal system.

"What about the second theorem?"

"It's about consistency. But if we jump ahead to that one I'm afraid you'll overload. We'll save it for another day."

"Good! I'm pretty much in the red zone already!"

7.2.3 Arithmetic

"So let's talk about the method that Gödel used in his first theorem," Miruka said. "Like so many other things we've talked about, it's a trip between two worlds."

She paused to look straight at Tetra and me in turn, as if to be sure she had our attention.

"Doing things like adding and multiplying natural numbers is called arithmetic. If you can set up a formal system that allows you to do that—in other words, a formal system of arithmetic—then you've formally defined the addition and multiplication of numbers. And if all that goes well, then you've found a way to express all possible arithmetic propositions using statements from your system. Things

like '2 is a prime number,' or 'the 17th power of 5 is 762939453125'."
She paused, and a wicked grin spread across her face. "Or have you?

"Let's go back and think one more time about just what a formal
system is. At its heart, it's all about *symbols*. In a formal system
of propositional logic, you line up symbols like $\boxed{\neg}$ and $\boxed{(}$ and
$\boxed{A}$ and $\boxed{)}$ to build up propositional formulas. If you're dealing
with a formal system of arithmetic, you'll use symbols like $\boxed{\neg}$ and
$\boxed{(}$ and $\boxed{x}$ and $\boxed{<}$ and $\boxed{y}$ and $\boxed{)}$. But there's nothing special
about those particular symbols. So long as we can tell the difference
between them, they could be anything, right?"

I was in unfamiliar territory, with no idea where we were headed.
I gave a slight nod, the best answer I could come up with.

"Well then," Miruka continued, "how about if in our formal system of arithmetic, we use *natural numbers* in place of these symbols.
Something like this."

· Use 3 in place of $\boxed{\neg}$

· Use 5 in place of $\boxed{(}$

· Use 17 in place of $\boxed{x}$

· Use 7 in place of $\boxed{<}$

· Use 19 in place of $\boxed{y}$

· Use 9 in place of $\boxed{)}$

Tetra furrowed her brow.

"A 3 instead of $\boxed{\neg}$? But . . . *why?*"

"Just as an example," Miruka said, still smiling. "I picked the
numbers at random."

"Yeah, but why numbers in place of the symbols?" I asked.

"Because we can use numbers in arithmetic."

"So you're doing it . . . just because you can?"

"You don't see where I'm going with this, I guess?"

I shook my head. "Not in the slightest."

"We can write formal systems using symbols. We can represent
symbols using natural numbers. We can manipulate natural numbers

using arithmetic. If you put all those together, that means we can use arithmetic to manipulate formal systems."

Miruka pushed her glasses up her nose.

"It all seems very natural to me," she said. "Of course, that's probably only because I live in a post-Gödel world."

I just sat there with my mouth half open. Beside me, Tetra groaned.

"This is absolutely, positively the *weirdest* math I've ever seen," she said.

7.2.4 Formal Systems of Formal Systems

Miruka was on a roll, though, and pressed on relentlessly.

"Let's talk Gödel numbers," she said. "So we've agreed to represent the symbols $\boxed{\neg}$, $\boxed{(}$, $\boxed{x}$, $\boxed{<}$, $\boxed{y}$, $\boxed{)}$ by the natural numbers $3, 5, 17, 7, 19$ and 9, respectively. Again, there's nothing special about these numbers.

"Since we can just think of the propositional formula $\boxed{\neg}\boxed{(}\boxed{x}\boxed{<}\boxed{y}\boxed{)}$ as a string of symbols, we can also represent it as the sequence $\langle 3, 5, 17, 7, 19, 9 \rangle$. But then we can also use prime exponents to change this sequence of numbers into a single natural number."

"Huh," I grunted. I still wasn't sure where she was headed, but it was starting to look pretty cool.

"What does that mean, exactly?" Tetra asked.

"Look what we've got right now," Miruka said.

"I want to change this sequence of numbers into a single number. A good way to do that is to line up the primes in increasing order..."

$$\langle 2, 3, 5, 7, 11, 13, \dots \rangle$$

"... and use my sequence as exponents on these primes. Then I'll take the product of the whole lot to create some big number."

$$2^{\boxed{\neg}} \times 3^{\boxed{(}} \times 5^{\boxed{x}} \times 7^{\boxed{\le}} \times 11^{\boxed{y}} \times 13^{\boxed{)}}$$

$$= \quad 2^3 \times 3^5 \times 5^{17} \times 7^7 \times 11^{19} \times 13^9$$

$$= \quad 8 \times 243 \times 762939453125 \times 823543$$

$$\times 61159090448414546291 \times 10604499373$$

$$= \quad 79217987141081571017188492699098480419873046875000$$

"I can do the same thing to any propositional formula to generate its own unique number—its Gödel number."

"That's ... the biggest number I've ever seen," Tetra said, her eyes wide.

"Big isn't interesting," Miruka said. "What's interesting is that it's a unique number for this propositional formula. Even more interesting is that we can do the same thing to formal proofs. After all, a formal proof is just a finite sequence of propositional formulas. And since we can represent a propositional formula as a number, we can represent a formal proof as a finite sequence of natural numbers. Play the same trick twice, and it all crunches down into a single natural number. A big one, maybe, but a number all the same."

A finite sequence of finite sequences of natural
numbers
↓
A finite sequence of natural numbers
↓
A natural number

"Proof to number in two easy steps."

"That's just nuts," I said, shaking my head.

"It's just arithmetic." Miruka smiled. "Oh, but it gets better. We used products of primes to generate our Gödel numbers. Why is that convenient?"

I had to think for a moment, but prime number fan that I am, realization soon dawned on me.

"Prime factorization! You can go backwards!"

Miruka nodded.

"That's right. Given a Gödel number, we can reconstruct the propositional formula—or formulas—that generated it, using prime factorization to generate the associated sequence of numbers. Of course, if you're going to play tricks like that you need some way of making sure that the sequence of numbers you get is a valid propositional formula. A predicate that serves as a propositional formula detector, if you will."

"What do you mean, a detector?" Tetra asked.

"A statement that would be true when you fed it this huge number, for example, since this is the Gödel number for a valid formula. You can create other interesting things, too. Like an axiom detector—a predicate that tells you if a given number is the Gödel number for a valid axiom. Even a proof detector that takes two numbers, Gödel numbers for a formal proof and for some statement, and tells you if the proof supports the statement."

"Math that checks itself," I muttered, doing my best to grasp the implications of what Miruka was describing. "But is all that actually possible?"

"Not only is it possible, Gödel gives concrete examples in his proof of the incompleteness theorems."

"Whoa."

"He goes even further than that, though. Can I interest you in a provability detector?"

"Is that exactly what it sounds like?"

"If it sounds like a predicate that tells you if a given number is the Gödel number for some statement, and furthermore whether a formal proof of that statement exists, then yes."

"You've got to be joking."

"Well, to be fair, that one's a bit different from the others, so calling it a 'detector' along with the rest might be pushing things a bit. But still."

Miruka leaned back in her chair, hands behind her head as she looked up at the ceiling.

"Let's take a step back and look at what's going on here," she said. "We're saying we can create things like propositional formula detectors and proof detectors, which allow us to represent formal systems as arithmetic. But we started out saying we could create a formal system *for* arithmetic. So we've represented a formal system using arithmetic, and arithmetic as a formal system. When you combine the two, what do you end up with?"

"A formal system of formal systems," I said, my voice barely a whisper.

Formal systems
(propositional formulas, proofs, ...)
↓
Arithmetic
(propositional formula detectors, proof detectors, ...)
↓
Formal systems
(propositional formulas that describe propositional formula detectors, propositional formulas that describe proof detectors, ...)

7.2.5 Some Vocabulary

Tetra threw both hands up in the air. She looked exhausted.

"Stop, stop," she said. "I don't think I can take any more!"

"Hmph," Miruka said. "Too many words again?"

"So many, I'm starting to feel dizzy."

"Maybe a vocabulary list would help," I said.

Miruka began writing out a quick list of words in the notebook. "Something like this?"

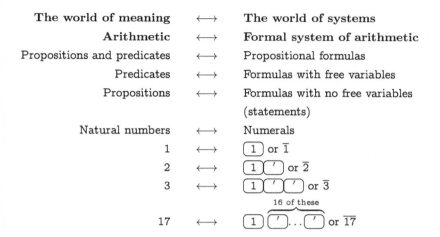

"Better," Tetra said. "I think."

7.2.6 Numerals

"Er, actually no," Tetra said. "What do the lines like this mean?"

She pointed to a line Miruka had written: '1 ′ ′ or 3̄.'

I was impressed that she kept asking questions.

"Numerals are how you represent the concept of natural numbers in the formal world," Miruka said. "The prime marks are from the idea of successors in the Peano axioms. We formally represent the natural number 3 as the numeral 1 ′ ′ , a sequence of three symbols. That's too much of a pain when you start dealing with bigger numbers, though, so we're going to allow a shorthand, writing the natural number with a line over it. That lets us write our proposition that 17 is prime like this."

"Ah, okay," Tetra said.

"Obtaining the numeral 1 ′ ′ from the natural number 3 is an example of taking something from the meaning world to the formal world. Going the other way, converting a propositional formula like $\boxed{\neg}\,\boxed{(}\,\boxed{(}\,\boxed{x}\,\boxed{<}\,\boxed{y}\,\boxed{)}\,\boxed{)}$ into a Gödel number is like scooping a concept from the formal world and bringing it into the meaning world, in the form of a natural number."

"Neat!" Tetra said. "It's kind of like somebody from the real world appearing in a novel, or a fictional character appearing in the real world!"

7.2.7   Diagonalization Revisited

"Let's get back to diagonalization," Miruka said. "I'm going to use a new vocabulary word too: 'univariate propositional formula.' It's a mouthful, but all it means is a propositional formula that only has one free variable. Take this, for example."

$$\boxed{\forall}\,\boxed{m}\,\boxed{\forall}\,\boxed{n}\,\boxed{(}\,\boxed{(}\,\boxed{m}\,\boxed{<}\,\boxed{x}\,\boxed{\wedge}\,\boxed{n}\,\boxed{<}\,\boxed{x}\,\boxed{)}\,\boxed{\rightarrow}\,\boxed{m}$$
$$\boxed{\times}\,\boxed{n}\,\boxed{\neq}\,\boxed{x}\,\boxed{)}$$

"This is a univariate propositional formula with the free variable x. Let's call it f. Of course, a univariate propositional formula is just a kind of propositional formula, so we can find a Gödel number for it. Let's assume that the Gödel number we found for f was 123."

"Sure is a lot smaller than the last one we found!" Tetra said.

"Its true Gödel number would be much larger, but this is good enough for the sake of demonstration."

"Gotcha."

"Anyway, 123 is a natural number, a concept from arithmetic in the world of meaning. Let's take it to the formal world, by creating the numeral for 123.

122 of these

"Of course, we want to write that using our shorthand version, $\overline{123}$. Now, let's replace all instances of the free variable x in our univariate propositional formula f with the numeral $\overline{123}$. Then we get a propositional formula like this."

$$\boxed{\forall}\,\boxed{m}\,\boxed{\forall}\,\boxed{n}\,\boxed{(}\,\boxed{(}\,\boxed{m}\,\boxed{<}\,\overline{123}\,\boxed{\wedge}\,\boxed{n}\,\boxed{<}\,\overline{123}\,\boxed{)}\,\boxed{\rightarrow}\,\boxed{m}$$
$$\boxed{\times}\,\boxed{n}\,\boxed{\neq}\,\overline{123}\,\boxed{)}$$

"This updated propositional formula doesn't have a free variable like f did, so it's a statement. Let's use this notation to denote a statement like that."

$$f\langle \overline{f} \rangle$$

"Let me make sure I've got this right," Tetra said. "This $f\langle \overline{f} \rangle$ is a statement, one that we created from a formula f by replacing its free variable with a number?"

"Not just any number, though," Miruka said. "Remember, we replaced x with the Gödel number of f."

"Oh, right! Okay, I'm good."

"One more vocabulary word, then. I want to call an $f\langle \overline{f} \rangle$ like this a 'diagonalization of f.'"

Diagonalization

f	A univariate propositional formula
$f\langle \overline{f} \rangle$	A statement created by replacing all instances of the free variable in f with the Gödel number of f

---

"I kinda see what's going on here," I said, "and I kinda don't."

"Speaking in terms of formal operations tends to cloud things," Miruka said. "Maybe an analogy using natural language will make it clearer. Try thinking of creating $f\langle \overline{f} \rangle$ from f as starting with this sentence..."

x is some value

"...and changing it into this."

"x is some value" is some value

"Ah, I get this," I said. "Or like starting with this..."

x has twenty characters.

"...and changing it into this."

"x has twenty characters" has twenty characters.

"Or starting with this ..." Tetra said.

x isn't written in Esperanto.

" ... and rewriting it as this!"

"x isn't written in Esperanto" isn't written in Esperanto.

"Excellent examples all," Miruka said.

"So is all this somehow related to Gödel's incompleteness theorems?" I asked.

"Oh yes," Miruka said, nodding slowly. "Gödel created a specific
example of a formal system of formal systems, and used it to prove
his first incompleteness theorem. In that proof, he diagonalized a
statement like this."

There exists no formal proof of the diagonalization of x.

"What would that diagonalization look like?" Miruka asked, looking at Tetra.

"Um ... something like this?"

There exists no formal proof of the diagonalization of "There exists no formal proof of the diagonalization of x."

"Exactly. And Gödel proved the first incompleteness theorem by
investigating whether a formal proof of this statement exists. It's a
fun ride, with lots to see. Creating proof detectors out of prime numbers, self-referential statements through diagonalization ... Good
stuff. It makes me wonder what Gödel would have come up with
if he had become a composer or an architect or a programmer, instead of a mathematician."

Miruka looked off into the distance and sighed.

7.2.8   Setting Things Straight

Miruka's lecture was done, allowing Tetra and me to finally relax. Tetra had sunk back into her chair and was taking deep, slow breaths.

"Yuri would love this stuff," I said, rolling stiff shoulders.

"Speaking of which, why isn't she here?" Miruka asked.

"Who, Yuri? Uh, because this is a high school, ya think?"

"Our schools are too restrictive," Miruka grumbled. "Yuri needs to be set straight. She's been poisoned by the popular misconception that Gödel's incompleteness theorems are about the 'limits of reasoning,' or some such nonsense. It's a theory about *math*, and math alone. It's a rich theory, and one that we can learn a lot from. I want her to see that."

Miruka frowned and pushed my notebook back at me. She stood and began gathering her things.

"There's more," she said, "but I don't want to get into it without Yuri."

"Somewhere other than here, I hope."

"I know a much better place," Miruka said. "Narabikura Library."

7.3   Search for a Search

7.3.1   First Date

Just a few days later, Miruka and I went on our first date. Or whatever it was, given that it had been arranged via my mother.

"You be a gentleman, now," Mom said. "Mind your manners. And get her home before too late. You should walk her home, you know. And don't forget—"

I nodded mindlessly at my mother's endless litany of cautions and admonitions and somehow made it out the door.

"This way," Miruka said soon after we arrived at the amusement park, pulling me toward a display table covered in Lego blocks.

"I think that's supposed to be for little—okay, okay. I'm coming."

I had to admit, it was fun. I'd always had a weakness for Legos. As I built a three-dimensional Sierpinski gasket and Miruka worked on a Klein bottle, I wondered if this was the kind of thing one did on a date. Being surrounded by first and second graders didn't set quite the mood I'd imagined.

When we were done we bought some ice cream—vanilla for me, chocolate for Miruka—and sat across from each other to eat them.

"So, why did you call my mom, instead of me?" I asked.

"No reason," Miruka said. She pointed at my cone. "Gimme a bite."

"Uh . . . sure."

She took an experimental lick. She must have liked it, because before I knew it half of my ice cream had vanished.

"A bite, huh?" I said, laughing.

She was smiling, too. I thought back on our math talk the other day. She always looked happy when she was pulling apart complex mathematics, but this was different somehow. Like I was seeing a new side of her.

"What are you staring at?" she asked.

"Nothing. Just glad you're having fun."

"You're the fun one here. It's why everybody likes you."

"Yeah, right. Like I have half the friends you do," I said. "Even my mother thinks you're great. I think she's afraid I'm going to scare you off today."

Miruka made a typically undecipherable face and focused on finishing her ice cream. I looked around for something to change the conversation.

"Hey, let's go ride that," I said, pointing behind her.

Miruka turned and stared for a moment at the ferris wheel, watching it spin against the clear blue sky.

"If you want," she said.

We bought our tickets and stood in line behind another couple, college students by the look of them. The boy whispered something in his date's ear, causing her to giggle and jab his back. I looked away and watched the colorful gondolas as they sank and rose again.

After a brief wait the couple in front of us finally clambered into a bright orange cabin marked #16. I looked back and saw that ours would be the light blue #17.

A lovely prime . . .

When it was in place, the attendant opened the door and gestured us in. Miruka mumbled something as she climbed up.

"Did you say something?" I asked.

"Just hurry up and sit down."

I sat in the seat across from her just as the lock clicked.

I hadn't been in a ferris wheel in years. I looked through the small window, watching the complex geometric patterns that developed, then melted away as supporting wires moved against the stationary poles. I looked down and was surprised at how small everyone already looked.

I turned to Miruka, intending to make some lame joke about miniature worlds, but froze when I saw her. She had closed her eyes and looked ready to pass out.

"Are you okay?" I asked.

"I'm fine," she said. "Just . . . don't move."

"No, seriously."

I shuffled to her side of the cabin and sat next to her, causing the gondola to lurch.

"I said don't move! If this thing falls, we're dead!"

"Sorry, I'll move back . . . "

"No! You'll make it swing again!" Miruka wrapped her arms around me and put her head on my chest. "Just *don't move!* "

"Miruka, I—"

"Don't move," she repeated. "Just . . . don't move."

Miruka's hands clutched at my shirt. She pressed her head tightly against my chest, and I looked down at the black rolling sea of her hair. The citrus scent of her enveloped us.

"There are things I'm afraid of," she said. "Sometimes I get scared."

"If you're afraid of heights, you should have said—"

"I'm not talking about the ferris wheel."

I put my hand on the back of her head. Her hair felt like silk.

"It's okay, Miruka," I said. "You're okay."

I was worried she'd get angry with me again, but instead she exhaled and softened. We remained that way as we traced out our orbit, Miruka in my arms as I stroked her hair.

"Which do you think the prince enjoyed more?" Miruka whispered.

A gust of wind whistled through the wires outside the window, causing the cabin to creak.

"Finding the glass slipper, or finding the girl?"

Let us move on to the real numbers $\mathbb{R}$. Are they still countable? No, they are not, and the means by which this is shown—Cantor's diagonalization method—is not only of fundamental importance for all of set theory, but certainly belongs in The Book as a rare stroke of genius.

M. AIGNER & G. ZIEGLER
Proofs from THE BOOK

CHAPTER 8

Between Two Solitudes

> The two separate worlds or the two solitudes will surely have more to give each other than when each was a meager half.
>
> ANNE MORROW LINDBERGH
> *Gift from the Sea*

8.1 OVERLAPPING PAIRS

8.1.1 *What Tetra Noticed*

The March winds carried the scent of spring. The school was abuzz over the graduation ceremony the following day, but since I wasn't graduating yet, in the end it was just a day like any other. So like any other day, I headed to the library after my last class. When I arrived, I found Tetra already hard at work on a math problem.

"You're here early," I said.

Tetra looked up from her notebook and beamed.

"Hey! Yeah, I couldn't wait to get started on something."

"Is that another card from Mr. Muraki?" I asked, sitting down next to her.

"Sure is!"

Overlapping Pairs

Call two associated natural numbers a *pair*.

$$\langle a, b \rangle \qquad \text{paired natural numbers } a \text{ and } b$$

If $a + d = b + c$ for two pairs $\langle a, b \rangle$ and $\langle c, d \rangle$, then we say that these pairs *overlap*, which we express as $\langle a, b \rangle \overset{\circ}{=} \langle c, d \rangle$.

$$a + d = b + c$$
$$\Longleftrightarrow \quad \langle a, b \rangle \text{ overlaps } \langle c, d \rangle$$
$$\Longleftrightarrow \quad \langle a, b \rangle \overset{\circ}{=} \langle c, d \rangle$$

"It's a real weird one," Tetra said.

"Yeah, it isn't even a problem," I said, laughing.

"Another one of his research topics, I guess? One of those come-up-with-the-problem-yourself things?"

"Looks that way. There's something interesting about this $a+d = b + c$, I'd think. Something that isn't immediately obvious."

"Hmm. Y'know, when I first saw this card I was like, that's it? I mean, a pair of numbers, and simple addition? What could be so hard about it? There's no Greek letters, no limits to infinity... But when I tried to pull it all together, I realized I have no idea what's going on here. Total understanding of the parts, zero understanding of the whole."

I nodded. "Yep, I've been there."

"I'm still giving it my best shot, though! I'm doing like you suggested, going step by step until I run into the edge of my understanding."

"What are you doing in particular?"

"Well, I started out by making some of my own example pairs, first by letting a be 1, and $b = 1, 2, 3, \ldots$ Then I got this."

Tetra pointed to a line in her notebook.

$$\langle 1, 1 \rangle, \langle 1, 2 \rangle, \langle 1, 3 \rangle, \ldots$$

"Then I did the same thing with $a = 2$."

$$\langle\, 2, 1\,\rangle, \langle\, 2, 2\,\rangle, \langle\, 2, 3\,\rangle, \ldots$$

"Then I tried using all kinds of values."

$$\langle\, 12, 345\,\rangle, \langle\, 1000, 100000\,\rangle, \langle\, 314159, 265\,\rangle, \ldots$$

"Did you discover anything?" I asked.

"I did think about how sticking to the natural numbers means 0 will never show up, how you'll never get something like $\langle\, 0, 0\,\rangle$ or $\langle\, 0, 123\,\rangle$ or $\langle\, 314, 0\,\rangle$. Actually I was kind of surprised when I found myself seeing conditions like that. You know how I can be."

"See? I told you you're getting better at this."

"Yeah, well..." Tetra blushed. "So anyway, I noticed how *you* notice things like that when you make examples. Hey, I guess that's meta-notice!"

I feigned a groan.

"Did you try anything else?" I asked.

"Oh, yeah! After that I wondered what the set of all pairs would look like. A set where all my examples would be elements, like this."

$$\{\, \langle\, 1, 1\,\rangle, \langle\, 1, 2\,\rangle, \langle\, 1, 3\,\rangle, \ldots,$$
$$\langle\, 2, 1\,\rangle, \langle\, 2, 2\,\rangle, \langle\, 2, 3\,\rangle, \ldots,$$
$$\langle\, 12, 345\,\rangle, \langle\, 1000, 100000\,\rangle, \langle\, 314159, 265\,\rangle, \ldots\}$$

"Not that I learned much from doing that, though," she frowned.

"What were you working on just now?"

"Trying to figure out this 'overlapping' stuff that's on the card, and what $\langle\, a, b\,\rangle \overset{\circ}{=} \langle\, c, d\,\rangle$ means."

Tetra paused and bit her lower lip.

"No, that's not quite it," she said. "I guess this is what I really don't understand."

$$a + d = b + c$$

"No?" I said.

"Umm... I'm still not saying it right. I guess the real edge of my understanding is what this equation really *means*. Like, on the surface it's just a simple addition problem, so my initial reaction was, 'so what'? Sure, it's the condition for one pair to overlap another pair, I get that. But that can't be all. It seems like I'm supposed to go beyond that, but I can't seem to. It's like I've hit some kind of invisible force field that's preventing me from going any further."

Tetra pantomimed banging on an unseen wall.

"Wow, Tetra," I said. "That's... You're pretty amazing."

Tetra grimaced. "What's so amazing about not getting anywhere?"

"Knowing there's somewhere to get to, and not giving up on finding it," I said. "Sometimes it takes you a while to figure stuff out, but once you do it's yours for keeps. That's a good talent to have."

"I'd rather have the talent of figuring stuff out."

"Well, let's give that a shot. I don't know what's so interesting about this card either, not yet. But maybe with some more examples we can find something."

"More examples?"

"Sure. Here's what you're banging up against, right?"

$$\langle a, b \rangle \overset{\circ}{=} \langle c, d \rangle \iff a + d = b + c$$

"Yeah, I guess..."

"So how about creating some overlapping pairs that fit this, and see what you find."

"Oh, right! I guess I haven't done that yet, have I. Okay, give me a minute!"

"You bet."

Tetra went serious as she returned to her notebook. I watched her work for a while, enjoying the play of expressions that crossed her face: Furrowed eyebrows when she made mistakes. Wide eyes when she discovered something new. Tilted head as she pondered where to go next. Surreptitious glances my way, probably wondering if she should ask me something. A slight head shake when she decided against it.

My mind wandered, and I started thinking about the upcoming year. I would have college entrance exams looming over my head, which seemed like a pointless bother, a distraction from the things

I really wanted to do. Elementary and middle school perversely felt more attractive, if only because they were free of such silly burdens. The irony was that my good grades back then had gotten me into this high school, which placed a heavy emphasis on getting students into first-rate colleges... bringing me back to the pressures of entrance exams.

"Okay, I've got some," Tetra said, pushing her notebook my way and rousing me from my brooding. "The card says the condition for overlapping is that $\langle a, b \rangle \overset{\circ}{=} \langle c, d \rangle \iff a + d = b + c$, so I went looking for sets of four numbers that make that work. Like when a and b are 1, and c and d are 2, for example."

$$\langle 1, 1 \rangle \overset{\circ}{=} \langle 2, 2 \rangle$$

"Sure, that works," I said.

"Here's another one, since $1 + 3 = 2 + 2$."

$$\langle 1, 2 \rangle \overset{\circ}{=} \langle 2, 3 \rangle$$

"Huh. I guess once you see the trick, you could make lots of these."

"Yeah! It gets easier when you forget about the letters and think of adding 'outside numbers' and 'inside numbers' instead. Like this."

$$\langle \text{ⓐ}, b \rangle \quad \text{and} \quad \langle c, \text{ⓓ} \rangle$$

"See? We're adding numbers on the outside here. When we add $b + c$, we're adding numbers on the inside."

$$\langle a, \text{ⓑ} \rangle \quad \text{and} \quad \langle \text{ⓒ}, d \rangle$$

Tetra scrunched her face.

"Of course, I guess that's another 'so what?' kinda thing to notice."

"Maybe, maybe not. You never know what's going to end up being useful."

"Oh, I also noticed that this reminds me of comparing ratios. You know how you can check for similar ratios like $2 : 3$ and $4 :$

6 by multiplying outside and inside numbers and comparing the products?"

$$\underset{\text{inner}}{\overset{\text{outer}}{\overline{2}:\underline{3}}} \quad = \quad \underset{\text{inner}}{\overset{\text{outer}}{\underline{4}:\overline{6}}} \quad \Longleftrightarrow \quad \overset{\text{outer product}}{\overline{2}\times\overline{6}} \quad = \quad \underset{\text{inner product}}{\underline{3}\times\underline{4}}$$

"Well this kinda feels like the same thing," Tetra said. "You check if two pairs like $\langle 2,3 \rangle$ and $\langle 4,5 \rangle$ overlap by comparing their inner and outer sums."

$$\underset{\text{inner}}{\overset{\text{outer}}{\langle \overline{2},\underline{3} \rangle}} \quad \overset{\circ}{=} \quad \underset{\text{inner}}{\overset{\text{outer}}{\langle \underline{4},\overline{5} \rangle}} \quad \Longleftrightarrow \quad \overset{\text{outer sum}}{\overline{2+5}} \quad = \quad \underset{\text{inner sum}}{\underline{3+4}}$$

"Kinda similar, don't you think?" Tetra shrugged.
I nodded. "Sure, that's neat."
"Anyway, here are the overlapping pairs I've found so far."

a	b	c	d	$a+d$	$b+c$	overlapping pairs
1	1	1	1	2	2	$\langle 1,1 \rangle \overset{\circ}{=} \langle 1,1 \rangle$
1	1	2	2	3	3	$\langle 1,1 \rangle \overset{\circ}{=} \langle 2,2 \rangle$
1	2	2	3	4	4	$\langle 1,2 \rangle \overset{\circ}{=} \langle 2,3 \rangle$
1	3	2	4	5	5	$\langle 1,3 \rangle \overset{\circ}{=} \langle 2,4 \rangle$
2	1	3	2	4	4	$\langle 2,1 \rangle \overset{\circ}{=} \langle 3,2 \rangle$
3	1	4	2	5	5	$\langle 3,1 \rangle \overset{\circ}{=} \langle 4,2 \rangle$
2	2	3	3	5	5	$\langle 2,2 \rangle \overset{\circ}{=} \langle 3,3 \rangle$
2	3	4	5	7	7	$\langle 2,3 \rangle \overset{\circ}{=} \langle 4,5 \rangle$

"Well done," I said. "By the way, I think I might have noticed something too. Mind if I talk about it? There might be spoilers."
"Go ahead, spoil me!"

8.1.2 What I noticed

"The first thing I wanted to do when I saw this equation was move stuff around," I said.

$$a + d = b + c \qquad \text{original equation}$$
$$a + d - b = c \qquad \text{move b to the left}$$
$$a - b = c - d \qquad \text{move d to the right}$$

"Shifting things around like that gives you this."

$$a - b = c - d$$

"Expressed as pairs, it looks like this."

$$\langle a, b \rangle \overset{\circ}{=} \langle c, d \rangle \iff a - b = c - d$$

"Wait, what just happened?" Tetra asked. "Are you saying that two pairs overlap when their elements have the same differences?"

"Well, that's what the math says."

"Maybe I understand what's going on even less than I thought."

"Actually, I'm still right there with you. Looks like there's more to these pairs than meets the eye."

8.1.3 Getting Ready

Tetra and I decided to take a break and wandered over to the auditorium to see how preparations for the ceremony were progressing. It was a madhouse, with teachers and students bustling about, lining up chairs and arranging flowers on the stage.

"We're gonna be here forever," Ay-Ay said.

"Everything should be ready by tomorrow," Miruka said.

"It better be! That's when the ceremony is!"

Miruka and Ay-Ay had been recruited to play piano at graduation, making them de facto members of the preparation committee. That meant they'd be at school until everything was set up, so I'd given up on us all being able to walk to the station together.

"What are you playing?" I asked.

"*Auld Lang Syne*," Ay-Ay said.

"And the school anthem," Miruka added.

I nodded. "Same as last year, then."

"Except that we're—" Miruka began, but Ay-Ay poked her in the ribs and they both fell silent.

I wonder what that was all about?

8.2 AT HOME

8.2.1 My Own Math

When I finally scored some "me time" that night it was already past eleven o'clock. Despite the late hour I sat at my desk and gathered notebook, pencils, and erasers to hunker down with some math—my own math.

I recalled my first few months in high school, when I first met Mr. Muraki. I remembered how he had told me to find time each day to work on my own math. At the time I thought it was a pointless thing to say—I loved math and had been studying it on my own every day for years.

But high school had kept me busier than I'd expected. More classes, more homework, more tests. Ceremonies and school events and a hundred other things competed for my time. It wasn't long before I found myself actually struggling to make space for any math beyond what was required.

Good advice, Mr. Muraki.

8.2.2 Compressed Expressions

I pondered Mr. Muraki's odd problem, wondering where it was pointing. It didn't ask for a proof of an opaque identity or demand that I solve some gnarly equation. It just stated some simple definitions, then left me hanging.

A pair is two natural numbers, and if $a + d = b + c$ *for two pairs* $\langle a, b \rangle$ *and* $\langle c, d \rangle$, *then those pairs overlap...* Hmm.

Mr. Muraki's cards were usually designed to nudge us toward learning something new, but this one didn't give me much to go on. Not sure what else to do, I started jotting down random properties that came to mind. For example, I noticed that all pairs in the form $\langle a, a \rangle$, in other words pairs of the same number, would overlap.

$$\langle 1, 1 \rangle \overset{\circ}{=} \langle 2, 2 \rangle \overset{\circ}{=} \langle 3, 3 \rangle \overset{\circ}{=} \cdots$$

Proving that was easy enough: $m + n = m + n$ would hold for any natural numbers m, n, so $\langle m, m \rangle$ and $\langle n, n \rangle$ would overlap by definition.

$$\langle m, m \rangle \overset{\circ}{=} \langle n, n \rangle \iff m + n = m + n$$

Talking with Tetra I'd noticed how rearranging $a + d = b + c$ as $a - b = c - d$ showed that a pair $\langle \text{left}, \text{right} \rangle$ would overlap any other pair where the difference between 'left' and 'right' was the same. I created some examples of this, using pairs with differences of 1.

$$\langle 2, 1 \rangle \overset{\circ}{=} \langle 3, 2 \rangle \overset{\circ}{=} \langle 4, 3 \rangle \overset{\circ}{=} \cdots \qquad (\text{left} - \text{right} = 1)$$

I could also flip the order of those elements to create overlapping pairs with a difference of -1.

$$\langle 1, 2 \rangle \overset{\circ}{=} \langle 2, 3 \rangle \overset{\circ}{=} \langle 3, 4 \rangle \overset{\circ}{=} \cdots \qquad (\text{left} - \text{right} = -1)$$

But, to borrow Tetra's words, so what?

That made me recall Tetra's 'pretend you know nothing' approach to math, but it didn't seem applicable here. At the time we'd been talking about the Peano axioms, about how when you're creating successors you didn't want to think about what they *mean*, you just want to follow the axioms—doing so would give you the structure of the natural numbers just as a result of the restrictions imposed by the axioms.

Hang on... $a - b = c - d$ is a kind of restriction too, isn't it? I wonder if that means these pairs have some kind of structure to them. If you collected all the pairs with the same difference $a - b$ into a set, I wonder what that would look like.

I turned back to my notebook and stared at what I'd written.

$$\langle 1, 1 \rangle \overset{\circ}{=} \langle 2, 2 \rangle \overset{\circ}{=} \langle 3, 3 \rangle \overset{\circ}{=} \cdots$$

The set of pairs that overlap $\langle 1, 1 \rangle$ would look like this.

$$\{\langle 1, 1 \rangle, \langle 2, 2 \rangle, \langle 3, 3 \rangle, \ldots\}$$

So where is this pointing to?

"Examples hold compressed truths," Miruka had once said. We search for patterns in examples, and that leads us to concise descriptions.

So maybe I'll get somewhere by using an intensional definition of this set?

$$\{\langle 1,1 \rangle, \langle 2,2 \rangle, \langle 3,3 \rangle, \ldots\} = \{\langle a,b \rangle \mid a \in \mathbb{N} \land b \in \mathbb{N} \land a-b = 0\}$$

Meh, I know that $a \in \mathbb{N}$ and $b \in \mathbb{N}$ are conditions here. No need to explicitly write that.

$$\{\langle 1,1 \rangle, \langle 2,2 \rangle, \langle 3,3 \rangle, \ldots\} = \{\langle a,b \rangle \mid a-b = 0\}$$

Easy enough to do this with other sets too. Like, when the difference is 1 . . .

$$\{\langle 2,1 \rangle, \langle 3,2 \rangle, \langle 4,3 \rangle, \ldots\} = \{\langle a,b \rangle \mid a-b = 1\}$$

. . . or when it's −1.

$$\{\langle 1,2 \rangle, \langle 2,3 \rangle, \langle 3,4 \rangle, \ldots\} = \{\langle a,b \rangle \mid a-b = -1\}$$

That's shorter than giving specific elements, I guess. But I wonder if I can't go shorter still?

Inspiration slapped me upside the brain, startling me to the point where I stood up from my seat.

These pairs. Could they actually be . . . the integers?

I mentally took a step back and thought this through one more time to be sure.

They are! The set of pairs of natural numbers form the structure of the integers!

What had flashed into my mind was a one-to-one correspondence between integers n and sets of pairs with difference n.

$$\vdots$$

$$\{\langle 3,1 \rangle, \langle 4,2 \rangle, \langle 5,3 \rangle, \ldots\} \quad \longleftrightarrow \quad +2 \quad \text{difference} +2$$

$$\{\langle 2,1 \rangle, \langle 3,2 \rangle, \langle 4,3 \rangle, \ldots\} \quad \longleftrightarrow \quad +1 \quad \text{difference} +1$$

$$\{\langle 1,1 \rangle, \langle 2,2 \rangle, \langle 3,3 \rangle, \ldots\} \quad \longleftrightarrow \quad 0 \quad \text{difference} \quad 0$$

$$\{\langle 1,2 \rangle, \langle 2,3 \rangle, \langle 3,4 \rangle, \ldots\} \quad \longleftrightarrow \quad -1 \quad \text{difference} -1$$

$$\{\langle 1,3 \rangle, \langle 2,4 \rangle, \langle 3,5 \rangle, \ldots\} \quad \longleftrightarrow \quad -2 \quad \text{difference} -2$$

$$\vdots$$

Something beautiful was coalescing before me, but it hadn't taken full form. A simple correspondence wasn't enough. What I really wanted was some way to naturally view these sets of pairs as the integers themselves.

So what does it take to be an integer? If I'm an integer, what's my fundamental nature?

Too many ideas were crashing down on me at once, so I stopped, took a deep breath, and tried to collect myself.

I should start simple. With addition, say.

If these things were integers, it should be easy to add them. The $+$ operator was already being used for the natural numbers, so I took a stab at defining a $\overset{\circ}{+}$ operator for addition of pairs.

$$\langle 1,2 \rangle \overset{\circ}{+} \langle 2,3 \rangle \text{ equals} \ldots \text{ what?}$$

I didn't have any equations to work with or any convenient tidbits to dredge up from memory. I had to think of a way to define this new kind of addition.

Problem 8-1 (Adding pairs)

Give a definition of the $\overset{\circ}{+}$ operator for adding pairs $\langle a, b \rangle$ and $\langle c, d \rangle$.

8.2.3 Defining Addition

So how can I define adding pairs in a way that's similar to integer addition?

Another Miruka quote sprang to mind: "Isomorphic mappings are the root of meaning."

I was growing increasingly excited, like I was about to witness something being born, some hidden structure being uncovered. Something free, yet constrained. Constrained, and yet free...

This feeling... This is why I do math.

I shook my head to clear it and focus on the task at hand: where to begin to define addition of pairs.

I wanted to be able to view $\langle a, b \rangle$ and integer $a - b$ as the same thing, so maybe standard integer addition would be a good place to start.

Pair	$\longleftrightarrow$	Integer
$\langle a, b \rangle$	$\longleftrightarrow$	$a - b$
$\langle c, d \rangle$	$\longleftrightarrow$	$c - d$

First add the $a - b$ and $c - d$, then put things in left $-$ right form.

$$(a - b) + (c - d) = a - b + c - d \qquad \text{sums } a - b \text{ and } c - d$$
$$= (a + c) - (b + d) \quad \text{change to left} - \text{right form}$$

Okay, looks good. So that means we have $a + c$ on the left, and $b + d$ on the right. In other words, we want the sum of two pairs to look like this.

$$\langle \underline{a}, \underline{b} \rangle \overset{\circ}{+} \langle \underline{c}, \underline{d} \rangle = \langle \underline{a + c}, \underset{\sim}{b + d} \rangle$$

But that means I just have to add lefts and rights!

$$(a - b) + (c - d) = (a + c) - (b + d)$$

So if that's the case, how about just defining addition of pairs like this?

$$\langle a, b \rangle \overset{\circ}{+} \langle c, d \rangle = \langle a + c, b + d \rangle$$

Yeah, I think I'm heading in the right direction here. Let's give it a test drive with $1 + 2$. *A pair that corresponds with* 1 *would be, let's see,* $1 = 3 - 2$, *so I'll use* $\langle 3, 2 \rangle$. *And* $2 = 3 - 1$, *so I'll use* $\langle 3, 1 \rangle$ *for* 2. *Now let's add those up, using my definition.*

$$\langle 3, 2 \rangle \stackrel{\circ}{+} \langle 3, 1 \rangle = \langle 3 + 3, 2 + 1 \rangle$$
$$= \langle 6, 3 \rangle$$

Looks good! $6 - 3 = 3$, *so everything's working out!*

$$
\begin{array}{ccccc}
\langle 3, 2 \rangle & \stackrel{\circ}{+} & \langle 3, 1 \rangle & = & \langle 6, 3 \rangle \\
\updownarrow & \updownarrow & \updownarrow & \updownarrow & \updownarrow \\
1 & + & 2 & = & 3
\end{array}
$$

Hang on, hang on. I'm just stating the obvious here. This is similar to what I want to say, but I'm not quite there. Time to step back and clean things up. Shouldn't be too hard.

What I need is a comparison between integers and pairs, $=$ *and* $\stackrel{\circ}{=}$. *Sure,* $+$ *and* $\stackrel{\circ}{+}$ *too, but I think I've jumped the gun defining addition while even equality is such a mess. I never even defined what it means for two pairs* $\langle a, b \rangle$ *and* $\langle c, d \rangle$ *to be equal. I guess that would be something like this.*

$$\langle a, b \rangle = \langle c, d \rangle \iff (a = c \wedge b = d)$$

Okay, now to the important part. Having defined pair addition as $\langle a, b \rangle \stackrel{\circ}{+} \langle c, d \rangle = \langle a + c, b + d \rangle$, *if I have a pair X that overlaps* $\langle a, b \rangle$, *and a pair Y that overlaps* $\langle c, d \rangle$ *and a pair Z that overlaps* $\langle a + c, b + d \rangle$, *then I need this to be true:*

$$X \stackrel{\circ}{+} Y \stackrel{\circ}{=} Z$$

Being able to do that is what will make pairs feel like integers.

World of pairs	$\longleftrightarrow$	World of integers
Set of all pairs overlapping $\langle a, b \rangle$	$\longleftrightarrow$	Integer $a - b$
Pair addition ($\stackrel{\circ}{+}$)	$\longleftrightarrow$	Integer addition ($+$)
Pair equality ($\stackrel{\circ}{=}$)	$\longleftrightarrow$	Integer equality ($=$)

I became increasingly excited as I saw these two worlds matching up.

Answer 8-1 (Adding pairs)

Addition of two pairs $\langle a, b \rangle$ and $\langle c, d \rangle$ is defined by

$$\langle a, b \rangle \overset{\circ}{+} \langle c, d \rangle = \langle a + c, b + d \rangle.$$

I'll bet this means that zero corresponds with all pairs that overlap $\langle a, a \rangle$! Oh, and if I swap the position of the elements in $\langle a, b \rangle$, then what happens with $\langle b, a \rangle$?

There's so much here to look into!

8.2.4   In the Zone

Around two in the morning I stumbled down to the kitchen, where I drained a glass of water. I was exhausted, but I had managed to define how to handle switching parity, subtraction, and inequalities of pairs.

$$\overset{\circ}{-} \langle a, b \rangle = \langle b, a \rangle \qquad\qquad \text{switching pair parity}$$

$$\langle a, b \rangle \overset{\circ}{-} \langle c, d \rangle = \langle a, b \rangle \overset{\circ}{+} (\overset{\circ}{-} \langle c, d \rangle) \qquad \text{pair subtraction}$$

$$\langle a, b \rangle \overset{\circ}{<} \langle c, d \rangle \iff a + d < b + c \qquad \text{pair inequality}$$

I had defined the integers using pairs of natural numbers—developed a whole new world of numbers, in a way. Not bad for a night's work. The more I explored mathematics, the vaster, more interesting the worlds I discovered.

I watched a bead of water trickle down my glass and thought about Mr. Muraki with newfound respect. He always seemed to hit our Goldilocks zone, giving us problems that weren't too easy or too hard. I don't know how he managed it, but I was grateful to have him as a teacher.

Creating the integers was enough math for one night. I started getting ready for bed, a smile on my face the entire time. But as I turned off the light a thought began to niggle at me: Tetra's comment about how the sums of inner and outer pair elements would be equal. My body was already shutting down, though, and I fell asleep before I could give that much thought.

8.3 EQUIVALENCE RELATIONS

8.3.1 The Graduation Ceremony

At the graduation ceremony the next day, I watched the seniors march one-by-one across the auditorium stage to collect their diplomas. I wondered what I would feel like next year, when I would be up there myself.

Fatigue prevented deep introspection, and I was fighting yawns as our principal delivered solemn words to the new graduates and their families. Like all things it eventually ended, and my eyes snapped open when the master of ceremonies took back the microphone, causing it to wail. After some closing words the ceremony was finally over, and the graduates began filing out of the room to the strains of *Auld Lang Syne*, passing up the aisle toward the exit. We stood and sent them off with a round of applause.

Halfway through a monster yawn I felt something in the room change. I looked around and realized the applause had stopped, as had the queue of graduates. Everyone was looking at Miruka and Ay-Ay at the piano next to the stage.

Something was happening to *Auld Lang Syne*. It was transforming into a melody that I knew well: our school song. But no, *Auld Lang Syne* was still there—it was being juxtaposed, not replaced. Miruka and Ay-Ay had composed a mashup.

No way this can work, I thought, but they somehow managed to pull it off. Potentially clashing chords neatly slid by each other, and disjointed rhythms formed mutual support rather than opposition.

The effect was astonishing. I felt an emotion I didn't have a good name for, some weird synthesis of confusion and impatience, worry and peace, resentment and joy. Strangest of all were feelings of nostalgia for my school. Memories from the past two years flickered through my mind.

"One, one, two, three . . . "

"What if the radius is zero?"

"Done and done."

I'd never been a huge fan of formal schooling, but I had to admit, some good things had happened at this school. I'd met Miruka and Tetra here and experienced the joy of learning with friends, teaching friends, even being challenged by friends.

Yet time kept flowing by. Mathematics is timeless, but not we humans. We come together and fall apart, buffeted by the winds of time.

I hurried to wipe away a tear I felt welling in my eye, but saw there was no need—everyone around me was crying too.

Ah, the heck with it, I thought, and let my tears flow.

8.3.2   What Pairs Produce

After the graduation ceremony, Tetra and I headed for the library. I looked forward to plunging into math to clear my head.

I gave Tetra a rundown of what I'd studied the previous night, and she seemed impressed. I noticed that her eyes were still red and puffy from crying.

"The integers? Just from these simple pairs? Wow."

"Neat, huh?" I said. "Well, to be precise, it's not that the pairs themselves correspond with integers. Sets of overlapping pairs do."

"So there's an infinitely large set of elements for each integer?"

"Right. And you can think of operations on a pair $\langle a, b \rangle$ as an operation on the integer $a - b$, so—"

I froze, having detected a faint citrus scent. I spun around to see Miruka standing there.

"What's got you all skittish?" she asked.

8.3.3   From Natural Numbers to Integers

"Took you long enough to get here," I said.

"I got called into the principal's office," Miruka said. "Side effect of not letting them know we'd tinkered with the playlist."

"Uh oh," Tetra said. "Did you get in trouble?"

"We might have, if they hadn't enjoyed our performance." Miruka dropped her stuff in a chair. "Is that a Muraki card?"

I handed her the card and told her what I'd found.

"Hmph," Miruka said. "Correct, but boring."

I don't know why I keep expecting praise...

"But that's kinda it, right?" I asked. "Isn't it natural to equate sets of overlapping pairs with the integers?"

Miruka pursed her lips and adjusted her glasses.

"Why do you think Muraki gave this definition?" she asked.

$$\langle a, b \rangle \stackrel{\circ}{=} \langle c, d \rangle \iff a + d = b + c$$

"The one that you messed up by moving stuff all over the place," she added.

"Is there some kind of special meaning to this?" Tetra asked.

"Nothing too amazing," Miruka said. "Actually there's nothing wrong with the way he did things, given that we already know about the integers. But let's pretend for a minute that we'd never heard of them, and define them from scratch."

"Not associate them, but *define* them?"

"Well, we'd have to if we didn't already know about them. In that case, $a - b$ would be undefined, because we could end up with something like $2-3$, which would make no sense—its solution doesn't lie within the realm of the natural numbers. So using $a - b$ to define 'overlapping' would be out of the question."

"And thus the $+$ operator instead," I said, nodding. "The $a + d$ and the $b+c$ can never leak out from the natural numbers, so they're safe to use in the definition."

$$\langle a, b \rangle \stackrel{\circ}{=} \langle c, d \rangle \iff a + d = b + c$$

8.3.4 Graphing

"There's something else you missed," Miruka said, "and once again you missed it because you never graph stuff."

I winced. This wasn't the first time she'd scolded me for being too attached to equations.

"What's there to graph? They're just—"

"Are you telling me you don't look at this and immediately think 'vector addition'?"

$$\langle a, b \rangle \overset{\circ}{+} \langle c, d \rangle = \langle a + c, b + d \rangle$$

"Well, now that you mention it... Yeah, that's pretty much exactly vector addition, isn't it."

$$(a, b) + (c, d) = (a + c, b + d)$$

"Um, guys?" Tetra said. "I'm kinda falling behind here..."

Miruka grabbed my notebook and pencil, then sketched a quick lattice of points.

"Both a and b are natural numbers, so we only need the first quadrant," she said.

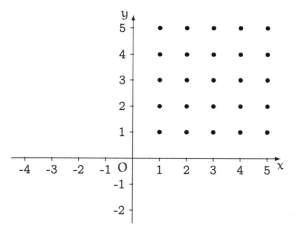

A first quadrant lattice

"Using a lattice like this we can think of x-coordinates as a and y-coordinates as b, then let the lattice points represent a vector (a, b), or a pair $\langle a, b \rangle$. So how can we show this 'overlapping' concept with respect to the set of these points? Maybe if we can do that, we'll see why Muraki called this 'overlapping' in the first place."

"I've been wondering that myself!" Tetra said.

"Then start by tying together overlapping pairs with lines, like this."

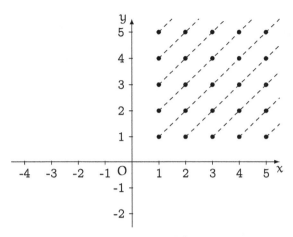

Drawing lines through overlapping pairs

"Whoa, so overlapping pairs are lined up in diagonals!" Tetra said.

"Ah, and that's where the overlapping comes from," I said, "how they line up in a two-dimensional coordinate plane. Each of these diagonal lines is associated with a set of overlapping pairs. With a single integer, in other words."

I filled in the integers to show the association.

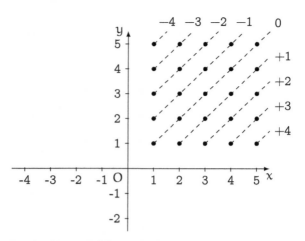

Each diagonal is associated with an integer

"Well, that's half of what you should do, at least." Miruka took the pencil from me and extended the dashed lines.

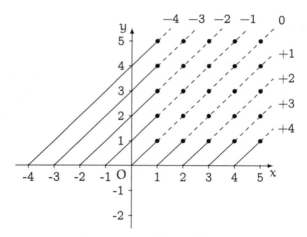

Extending the lines to their associated integer

"Too cool!" I said. "Extending the lines makes a projection onto the x-axis that corresponds to the associated integers! But hang on... If pair addition is like vector addition, does that mean you could graph a sum of ordinary vectors, and somehow see that sum as a projection?"

"Sure. Just do this diagonal projection of positional vectors on the lattice, and you'll get integer sums. Here's how you would show $\langle 1, 2 \rangle \overset{\circ}{+} \langle 4, 1 \rangle = \langle 5, 3 \rangle$."

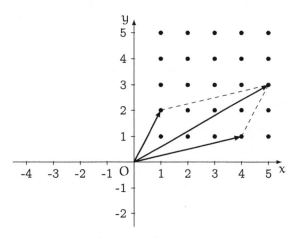

A sum of vectors

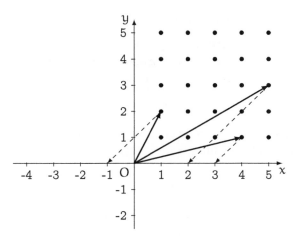

The projection gives a sum of integers

"See?" Miruka said. "$(-1) + 3 = 2$."

I was speechless.

You can view simple integer addition as a projection of two-dimensional vectors...

I hadn't digested this well enough to even begin considering the mathematical implications, but I could feel movement deep down, the snapping together of bonds between mathematical concepts I'd previously considered isolated from one another. And that wasn't just some hypothetical notion—here it was, right in front of me on paper. Miruka didn't leave much time for me to sit there stunned, though.

"Let's move on," she said.

8.3.5 Revisiting Equivalence Relations

"So," Miruka said. "Equivalence relations."

She tapped the notebook with her pencil several times, pausing either to consider how to proceed or for dramatic effect.

"Let S be the set of all pairs," she said.

$$S = \{\langle a, b \rangle \mid a \in \mathbb{N} \wedge b \in \mathbb{N}\}$$

"We're going to define a relation in S called 'overlapping,' like this."

$$\langle a, b \rangle \overset{\circ}{=} \langle c, d \rangle \iff a + d = b + c$$

"This relation is reflexive, symmetric, and transitive."

"Erm—" Tetra began, but Miruka cut her off with a nod.

"A reflexive relation is one like this."

$$\langle a, b \rangle \overset{\circ}{=} \langle a, b \rangle$$

"In other words, the relation holds when an element is compared with itself. It's like holding a mirror up to something and seeing the same thing reflected back.

"A symmetric relation is one like this."

$$\langle a, b \rangle \overset{\circ}{=} \langle c, d \rangle \overset{\text{implies}}{\Rightarrow} \langle c, d \rangle \overset{\circ}{=} \langle a, b \rangle$$

"In other words, the relation holds when you do a right–left swap of operands.

"A transitive relation is one like this."

$$\langle a, b \rangle \stackrel{\circ}{=} \langle c, d \rangle \stackrel{\text{and}}{\wedge} \langle c, d \rangle \stackrel{\circ}{=} \langle e, f \rangle \stackrel{\text{implies}}{\Rightarrow} \langle a, b \rangle \stackrel{\circ}{=} \langle e, f \rangle$$

"So if you can go from A to B, and from B to C, you can skip the interim B and go straight from A to C."

$$\underbrace{\langle a, b \rangle \stackrel{\circ}{=} \langle c, d \rangle}_{\text{from A to B}} \wedge \underbrace{\langle c, d \rangle \stackrel{\circ}{=} \langle e, f \rangle}_{\text{from B to C}} \Rightarrow \underbrace{\langle a, b \rangle \stackrel{\circ}{=} \langle e, f \rangle}_{\text{from A to C}}$$

"Reflexivity, symmetry, and transitivity are called equivalence rules, and a relation where each of these equivalence rules holds is called an equivalence relation. So our $\stackrel{\circ}{=}$ relation here is an example of an equivalence relation."

"Aren't those kind of obvious relationships, though?" Tetra asked. "Like, $=$ does the same thing, doesn't it?"

Miruka nodded.

"You're right that $=$ is an equivalence relation, yes. But that's not surprising, since equivalence relations were created as a generalization of equality. In fact, the best way to think of an equivalence relation is one where two elements are the same *in some way*." She pushed her glasses up her nose. "But let's look at some examples of operators that *don't* describe equivalence relations. The $<$ operator, for example. That's transitive, but not reflexive or symmetric."

NO	$a < a$	not reflexive
NO	$a < b \Rightarrow b < a$	not symmetric
YES	$a < b \wedge b < c \Rightarrow a < c$	transitive

"I see!" Tetra said.

"On the other hand, a $\leqslant$ relation is reflexive and transitive, but not symmetric."

YES	$a \leqslant a$	reflexive
NO	$a \leqslant b \Rightarrow b \leqslant a$	not symmetric
YES	$a \leqslant b \wedge b \leqslant c \Rightarrow a \leqslant c$	transitive

"Wait," Tetra said. "You can say $a \leqslant a$?"

"Sure. It means either $a < a$ *or* $a = a$, right?"

"Oh, of course."

"A question for you then," Miruka said. "What about the $\neq$ operator? Which of these properties does that have?"

"Well, it's the opposite of equals, so I'd think it doesn't have any of them."

Miruka shook her head. "Don't answer based on intuition. Go through them one by one."

NO	$a \neq a$	not reflexive
YES	$a \neq b \Rightarrow b \neq a$	symmetric
NO	$a \neq b \wedge b \neq c \Rightarrow a \neq c$	not transitive

"Oops, symmetric works, doesn't it," Tetra said.

"Hang on a sec," I said. "It's not so much that $\neq$ isn't transitive, it just isn't *always* transitive. Check it out."

with $a = 1, b = 2, c = 3 \ldots$ $1 \neq 2 \wedge 2 \neq 3 \Rightarrow 1 \neq 3$ transitive

with $a = 1, b = 2, c = 1 \ldots$ $1 \neq 2 \wedge 2 \neq 1 \Rightarrow 1 \neq 1$ not transitive

"Nice catch," Miruka said. "When I introduced these characteristics, I should have said 'for all elements.' In other words, if there's even a single counter-example, then the relationship doesn't hold."

"Um, Miruka?" Tetra nervously raised her hand. "I think I understand these relationships, but that thing you said about generalizing equality ... You can use it for numbers and sets and all kinds of stuff, so isn't it pretty general already?"

"An equivalence relation is one in which these three equivalence rules hold. Put another way, any relation that has each of these three properties is pretty much the same as equality. Identify the three characteristic properties of 'equals,' then find another relation that has the same properties—this 'overlapping pair' relation, for example. Since that's an equivalence relation, then anything you can say about equivalence relations will also apply to the overlapping pair relation."

"Wow, déjà vu," Tetra said. "Haven't we had this conversation before?"

"Pretty much. When we talked about group theory."

"That's right! When you said we can treat all operators that fulfill the group axioms as being the same thing!"[1]

"Break the 'equals' relation down into its basic parts, and extract the properties that characterize it. Then, create another relation that has the same properties. That's the fundamentals of analysis and synthesis. Do you see that?"

"Analysis and synthesis?"

"Analysis is picking things apart. Synthesis is putting them back together. When you analyze and synthesize things, you understand them better. They also become more interesting."

"So just what can you do with equivalence relations?" I asked.

"Exactly what you just did."

"Which would be?"

"Partitioning sets."

8.3.6   Quotient Sets

"Er, I partitioned sets?" I said.

"You did, using equivalence relations," Miruka said. "Think about it. There are infinitely many elements in the set S of all pairs. You used the equivalence relation $\overset{\circ}{=}$ to associate sets of overlapping pairs with the integers. Think of the set of all pairs as being like that first quadrant lattice, and think of the sets of overlapping pairs as being the diagonal lines. You partitioned the set of all lattice points into diagonal groups, right?"

"I guess, yeah?"

"When you partition a set using an equivalence relation, the result is a new set of sets called a 'quotient set,' and each of its elements is called an 'equivalence class.' So when you used $\overset{\circ}{=}$ to split up the set of all pairs, you got a quotient set with sets of overlapping pairs as its equivalence class elements. You write the quotient set like this."

$$S/\overset{\circ}{=}$$

"Kind of a weird notation, I know," Miruka said. "Just think of it like this."

$$\text{set/equivalence relation}$$

---

[1]See *Math Girls 2: Fermat's Last Theorem*, Ch. 6.

Tetra shook her head.

"I'm still not getting this. So what is this quotient set $S/\stackrel{\circ}{=}$? I'm not sure how I should be picturing it. I know they were diagonals in this graph, but what does that mean, mathematically?"

"I guess it's hard to get just from an explanation," Miruka said. "Okay, then. How about some extensional examples."

$$
S/\stackrel{\circ}{=} \;=\; \left\{
\begin{array}{ll}
\ldots, & \\
\{\langle 3,1 \rangle, \langle 4,2 \rangle, \langle 5,3 \rangle, \ldots\}, & \text{corresponds to } +2 \\
\{\langle 2,1 \rangle, \langle 3,2 \rangle, \langle 4,3 \rangle, \ldots\}, & \text{corresponds to } +1 \\
\{\langle 1,1 \rangle, \langle 2,2 \rangle, \langle 3,3 \rangle, \ldots\}, & \text{corresponds to } 0 \\
\{\langle 1,2 \rangle, \langle 2,3 \rangle, \langle 3,4 \rangle, \ldots\}, & \text{corresponds to } -1 \\
\{\langle 1,3 \rangle, \langle 2,4 \rangle, \langle 3,5 \rangle, \ldots\}, & \text{corresponds to } -2 \\
\ldots &
\end{array}
\right\}
$$

"I see," Tetra said. "So you're creating a set of sets."

"Equivalence classes, yes. And that's something you'll do often in certain areas of mathematics."

"Really?"

"Take the rationals, for example. Think of the set of all rational numbers as having elements that are number pairs created from numerators and denominators. You can partition that into equivalence classes using the equivalence relation 'has the same ratio.' "

"Hey! That's just what you said, Tetra!" I said.

"I said what now?"

"Remember? You were talking about how the inner and outer sums of pairs are the same, and how that reminded you of the inner and outer products of ratios."

"Umm..."

"So I guess the quotient set of the rational numbers would look something like this." I grabbed the notebook and started writing. "Lessee, we can write the pairs as $\langle\!\langle \text{numerator}, \text{denominator} \rangle\!\rangle$, so . . ."

$$
\left\{
\begin{array}{l}
\dots, \\[4pt]
\{\langle\!\langle +1,2 \rangle\!\rangle,\ \langle\!\langle +2,4 \rangle\!\rangle,\ \langle\!\langle +3,6 \rangle\!\rangle,\ \dots\ \},\ \text{associated with rational } \dfrac{+1}{2} \\[4pt]
\{\langle\!\langle +1,1 \rangle\!\rangle,\ \langle\!\langle +2,2 \rangle\!\rangle,\ \langle\!\langle +3,3 \rangle\!\rangle,\ \dots\ \},\ \text{associated with rational } +1 \\[4pt]
\{\langle\!\langle \ \ 0,1 \rangle\!\rangle,\ \langle\!\langle \ \ 0,2 \rangle\!\rangle,\ \langle\!\langle \ \ 0,3 \rangle\!\rangle,\ \dots\ \},\ \text{associated with rational } \ \ 0 \\[4pt]
\{\langle\!\langle -1,1 \rangle\!\rangle,\ \langle\!\langle -2,2 \rangle\!\rangle,\ \langle\!\langle -3,3 \rangle\!\rangle,\ \dots\ \},\ \text{associated with rational } -1 \\[4pt]
\{\langle\!\langle -1,2 \rangle\!\rangle,\ \langle\!\langle -2,4 \rangle\!\rangle,\ \langle\!\langle -3,6 \rangle\!\rangle,\ \dots\ \},\ \text{associated with rational } \dfrac{-1}{2} \\[4pt]
\dots
\end{array}
\right\}
$$

"Good job, Tetra," Miruka said.

"All I did was notice that these pairs and rational numbers kinda look the same."

"Noticing things like that is an important skill in mathematics. Things with similar forms often have similar natures."

"Partitioning into equivalence classes using equivalence relations, huh," I said. "This is pretty cool."

Miruka nodded.

"The results are cool too. When you partition this set by the equivalence relation 'has the same ratio,' the equivalence classes are sets of pairs with the same ratio. In other words, if you reduce each fraction, you'll get the same thing. It's a calculation that can move about in the set, so long as it doesn't jump bounds."

"You mean doesn't leak out of its partition, right?" Tetra asked.

"Right. But sometimes you'll want to pull out a single element from one of the equivalence classes to stand for its set as a whole. That's called a 'representative.'"

"Because it represents its class?"

"Right. After all, we're considering all the elements in a given equivalence class to be essentially the same thing."

"Why would we want to do that, though?"

"Because you might want to define something like a $\overset{\circ}{+}$ operator for adding elements within the quotient set. You need to be able to say that the result of that operator won't depend on how you chose the representative, in other words that the $\overset{\circ}{+}$ operation is well-defined."

"I was just thinking about that last night!" I said. "With regard to the number pairs, though."

"There are plenty of quotient sets other than the rationals," Miruka said. "For instance, you can partition the set of all integers by the equivalence relation 'has the same remainder after division by 3' to create the quotient set $\mathbb{Z}/3\mathbb{Z}$."

$$\mathbb{Z}/3\mathbb{Z} = \left\{ \begin{array}{l} \{\ldots, -6, -3, \quad 0, +3, +6, \ldots\}, \text{ remainder } 0 \\ \{\ldots, -5, -2, +1, +4, +7, \ldots\}, \text{ remainder } 1 \\ \{\ldots, -4, -1, +2, +5, +8, \ldots\} \text{ remainder } 2 \end{array} \right\}$$

"Why do you write that as $\mathbb{Z}/3\mathbb{Z}$?" Tetra asked.

"Think of the $3\mathbb{Z}$ part as meaning 'ignoring differences of multiples of 3.'" Miruka said.

"Okay. Can you give me another example of partitions?"

"Sure. How about taking the set of all students at this school and partitioning them by the equivalence relation 'is in the same grade.' Then you get a quotient set with equivalence classes comprising students in the same grade—the set of twelfth graders, the set of eleventh graders, and so on."

"Ooh! Ooh! I just noticed something!" Tetra said.

"What's that?" I asked, hoping for one of Tetra's unique flashes of mathematical insight.

"I *really* need to go to the bathroom," she said, and dashed off toward the exit.

8.4 AT THE RESTAURANT

8.4.1 *A Pair of Wings*

When I got home that evening, I was surprised by the absence of activity in the kitchen. I found my mother sitting in the living room, leafing through a magazine.

"Hey, Mom. What's for dinner?"

"I haven't decided. Your father says he'll be home late, that we should eat without him, but I can't think of anything good to make for two."

"Well, we haven't eaten out in a while."

"Hmm ... Italian?"

"Deal."

Thirty minutes later we were walking into one of our favorite restaurants, greeted by an enthusiastic "*Buona sera!*" from the staff. The irresistible odor of garlic, herbs, and olive oil made my stomach rumble. We were led to a booth decorated in red, white, and green—the food was far better than the decor—and immediately ordered our usuals, linguine pescatore for Mom, pizza margherita for me.

I looked around the restaurant while we waited for our food. The place was filling up with young couples and families. The background music was on the loud side, but not unpleasant—I'd always loved Italian guitar. That was soon drowned out, though, by most of the staff coming out to sing a birthday song to a guest at a nearby table.

"I can smell my pizza cooking," I said.

"You always did have such a good sense of smell," my mother said. "When the smell doesn't come from you, at least. Remember that time in kindergarten, when you—"

"Not that story again, please!"

"And to think, you're about to become a senior in high school. Where *does* the time go ... "

Mom propped her head on a hand and stared off into space.

Not that either, I thought. The sounds of guitar and laughing children faded away, and I began questioning myself again. Recent doubts about what I was doing and where I should be heading resurfaced, darkening my mood. I often heard adults refer to youth as a time of infinite possibilities, but the passage of time is one-dimensional. The pressure of needing to pick a direction and head that way felt almost too much to bear.

My mother glanced up from the dessert menu she'd been studying.

"You okay?"

"Just wondering what exactly I'm doing," I said.

"You're having a lovely dinner with your mother."

"Not that. It's just...I dunno. Sometimes it feels like I'm being pushed toward a cliff. School, entrance exams, college... It's like the ground is about to give out under me, and I'm not ready for it."

"So fly," Mom said.

"Fly?"

"Sure. If there's no ground left to walk on, spread your wings and fly. That's what the cliffs are for—to force you to take off and get moving under your own power. Are you scared?"

"Maybe. I've got good grades, but that doesn't seem enough somehow."

"Grades have nothing to do with it."

Mom started nudging a salt shaker with the tip of a finger, causing it to wobble its way across the table.

"Remember when you were first learning to walk?"

"Of course not. I was, what, one year old?"

"Well I do, like it was yesterday. You must have tripped and fallen and crashed into things a thousand times. But look at yourself now." She gave me a playful bop on the head with the menu. "Not ready for it. Ha! Don't worry—when you reach the cliff, I know you'll soar."

My mother's words didn't make any kind of logical sense at the time, but they were strangely reassuring nonetheless.

"I remember once when you were three, there was this terrible snowstorm..." Mom was dredging up another story I'd heard countless times before. I remained silent, however, and let her finish.

"Of course you'd pick that night to develop a hacking cough and break out into a fever. I honestly thought you were dying. Your father wasn't home, and the roads were shut down, so I bundled you up and carried you through the snow all the way to the emergency room, one town over. The whole thing was like—"

"—like something out of a disaster movie," I finished the story for her, knowing how it ended every time.

We were still laughing when the food arrived.

"Anyway, the point is, when push comes to shove you'll find a way to do what you have to do."

"I'll take your word for it," I said. "But for now, let's eat! I'm starved!"

I smothered my pizza in olive oil and chili peppers. I hadn't eaten anything so delicious in a long, long time.

8.4.2 Loosening Up

My mother had turned back to the dessert menu even as she was eating the last bites of her pasta.

"So I guess *torta al cioccolato* must be some kind of chocolate tart? But what's *crema catalana*? Something like *crème brûlée*?" She gave a critical sigh. "Dessert menus should always come with pictures. Just describing things with words isn't enough!"

"A lesson I keep having to relearn," I said, nodding.

"What you need is a big slice of relaxation pie," she continued.

"I didn't see that on the menu."

"Very funny. What I mean is that you need to learn to loosen up a little. Enjoy all those wonderful friends you have. How did you get so many cute girlfriends, anyway, shy thing that you are."

"What on earth are you going on about?"

"Hey, I know! We should all go on a drive somewhere. That would be so much fun!"

"Mom, please stop trying to arrange my social life."

"Let's see, we can fit five people in our car ... I've got to drive, so we'll put Miruka in the front seat, and Tetra and Ay-Ay can ride in back. Oh, and Yuri too! Perfect!"

"Five people? Wait, so I'm not invited!?"

Mathematics is the art of giving the same name to different things.

HENRI POINCARE
The Future of Mathematics

CHAPTER 9

Following Spirals

We started out on Earth, but we followed a spiral
and ended up here.

MOTO HAGIO
The Mosaic Spiral

9.1 $\frac{0}{3}\pi$ RADIANS

9.1.1 An Irritated Yuri

The following Saturday Yuri and I were seated at my dining room
table, eating rice crackers while my mother made tea. It was a laid-
back, peaceful day until my mother blurted out, "Oh, the other day,
when you were at the amusement park with Miruka—"

Yuri snapped up in her seat, her eyes darting back and forth
between my mother and me.

"Whoa whoa whoa, back up. What's this about you and Miruka
and an amusement park?"

"It's no big deal," I said, shrugging.

"No big—why didn't I hear about this?" she demanded. "Why
didn't you invite me? Wait a minute, are we talking just you and
Miruka . . . and that's it? As in you were out on a—"

"Fine, fine, next time it will just be you and me."

My mother slinked off into the kitchen, leaving me to face Yuri's
wrath alone.

"Oh, sure you will," Yuri spat. "Traitor!"

I'd seen Yuri like this before, so I gave up on trying to talk her down—getting her over the double insult of being denied both going to an amusement park and a chance to hang out with her idol would take some work. I ate the last bite of my cracker, then picked up my tea and took it to my room where I'd have the home field advantage. As I hoped, Yuri followed me despite her foul mood.

Once we got there she clammed up, though, and sat stewing in a cloud of indignation.

"Wanna do some math?" I asked, but her only response was a glare. "Fine. Be that way, then."

I turned to my desk to do some work, but Yuri refused to be ignored. For nearly twenty minutes she devised clever ways of distracting me, escalating from resentful sighs to "accidentally" bumping into my chair as she pulled a book off my bookshelf. I finally gave up on the mutual silent treatment and tried to appease her.

"Look, I'm sorry. I should have invited you to go with us."

Yuri's eyes flicked my way, and her expression softened a little.

"Hey, I just remembered. Miruka said she wants everyone to get together during spring break. You included."

That did it.

"For reals?" Yuri asked, eyeing me suspiciously.

"For reals. Something about wanting to teach you about Gödel's incompleteness theorems."

Yuri considered this for the briefest of moments.

"I'll be there," she said.

9.1.2 Trigonometric Functions

"You still owe me though," Yuri said. "You can start by teaching me about sines and cosines and all that."

"Trig functions? You're already studying that?"

"Not really. My teacher just mentioned something about sine curves, so I was wondering."

"Huh."

"I asked this friend of mine who's really good at math about them, but he doesn't know how to explain stuff like you do. We always end up in some kind of I-don't-get-it, why-don't-you-get-it fight, and I just walk away mad and not learning anything."

"Hmm... Well let's see if I can do any better."

I pulled out a notebook and turned to a fresh page.

"Start from the beginning," Yuri said.

"That would be the unit circle," I said, drawing one on the page before me. "That's a circle with radius 1, centered at the origin."

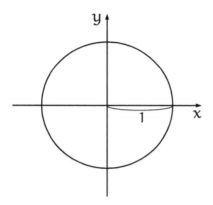

"Unit circle, right," Yuri said.

"Then we put a point P somewhere on that circle, and call the angle you need to get there θ."

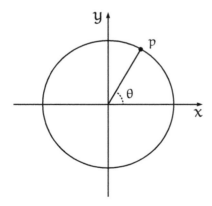

"What's a theta?"

"It's just a Greek letter, one we're using to name this angle. It's what you normally use to represent angles like this, but we can use something else if you prefer."

"Nah, theta it is. Carry on."

"Okay, do you see how angle $\angle\theta$ will change if we move point P around?"

"Sure."

"And the y-coordinate of point P will change along with it, right?"

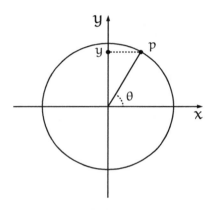

"Of course, because that's the height of P."

"Also, if we set θ to some specific angle, that determines what y will be, yeah?"

"Yep."

"Okay, well sometimes it's convenient to have some way to see what value y will be, depending on what $\angle\theta$ you get when point P moves around. In other words, a function that turns θ values into y values. That's called the sine function."

"Sine, as in all that sine–cosine stuff?"

"The very one. You write the function like this . . . "

$$y = \sin\theta$$

" . . . and you read it as 'y equals sine theta.' Plug a θ in, and you get a y-coordinate out."

"And that's it?"

"That's it."

9.1.3   sin 45°

"So what's the cosine function all about?" Yuri asked.

"Let's look at sine a little closer before we move on to that," I said. "For example, what's the value of $\sin\theta$ when θ is 0°?"

Yuri cocked her head and thought for a moment.

"Zero?"

"That's right. When θ is 0°, the y value is 0."

"Because P is just sitting there on the x-axis."

"Yep. So we can write this."

$$\sin 0° = 0$$

"I get it, I get it," Yuri said.

"Do you also see why y is 1 when θ is 90°?"

$$\sin 90° = 1$$

"Sure, because P is all the way up top."

"Good. So let's say we spin point P all the way around the circle, from 0° to 360°. What's the range in which $\sin\theta$—in other words the y-coordinate—will move?"

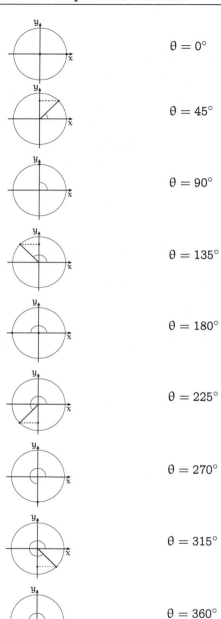

$\theta = 0°$

$\theta = 45°$

$\theta = 90°$

$\theta = 135°$

$\theta = 180°$

$\theta = 225°$

$\theta = 270°$

$\theta = 315°$

$\theta = 360°$

"I guess y goes from 0 to 1," Yuri said. "No, wait, −1 to 1!"

"You got it. So that means we can say this."

$$-1 \leqslant \sin\theta \leqslant 1$$

"In particular, it's −1 for $\sin 270°$, and 1 for $\sin 90°$."

"Yeah, yeah, I get it," Yuri said, annoyance creeping into her voice.

"Okay then, what's $\sin 45°$?"

"Well that's half of $\sin 90°$, so I assume it would be $\frac{1}{2}$."

"Bad assumption. Check out the graph."

"Huh? Hey, yeah... It looks like it's a little bit bigger."

"Gimme more than 'a little bit bigger.' I know you can find the exact value."

"How? Using a protractor or something?"

"Better to use calculations. Just think of this as being the diagonal of a square."

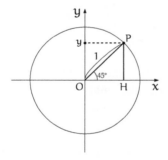

"So it's, uh, the length of a side of a square with a diagonal length of 1?"

"There ya go. So what's y?"

"Er... $\sqrt{2}$. No, $\frac{\sqrt{2}}{2}$."

"How'd you calculate that?"

"I didn't, I just know."

"Well, let me fill in the blanks anyway. You use the Pythagorean theorem, like this."

$$\overline{OH}^2 + \overline{PH}^2 = \overline{OP}^2 \qquad \text{from the Pythagorean theorem}$$

$$\overline{OH}^2 + \overline{PH}^2 = 1 \qquad \overline{OP} = 1, \text{ so its square is too}$$

$$\overline{OH}^2 + y^2 = 1 \qquad \text{because } \overline{PH} = y$$

$$y^2 + y^2 = 1 \qquad \text{because } \overline{OH} = y$$

$$2y^2 = 1 \qquad \text{add left side}$$

$$y^2 = \frac{1}{2} \qquad \text{divide both sides by 2}$$

$$y = \sqrt{\frac{1}{2}} \qquad \text{because } y > 0$$

$$= \sqrt{\frac{1 \times 2}{2 \times 2}} \qquad \text{multiply top and bottom by 2}$$

$$= \sqrt{\frac{1 \times 2}{2^2}} \qquad \text{denominator is a square}$$

$$= \frac{\sqrt{2}}{2} \qquad \text{remove } \sqrt{\ } \text{ from square}$$

"Maybe *you* use the Pythagorean theorem like that," Yuri said. "*I* just know that the diagonal of a square with side length 1 is $\sqrt{2}$. So if you want the diagonal to be 1 instead, you just divide everything by $\sqrt{2}$, which makes the square sides $\frac{1}{\sqrt{2}}$. Multiply the numerator and denominator to clean up, and there ya go—$\frac{\sqrt{2}}{2}$."

"Fair enough. By the way, $\sqrt{2}$ is approximately 1.4, right?"

"If you say so."

"Well 1.4 squared would be 1.96, which is pretty close, yeah?"

"Yeah, so?"

"So $\frac{\sqrt{2}}{2}$ would be approximately half of 1.4, or 0.7."

"I'll buy that, okay."

"Well, there's your answer. The value of $\sin 45°$ is around 0.7."

"Oh, sure. Right."

I dug around in my desk, pulled out a calculator, and did some tapping.

"A slightly more accurate value would be... $\sqrt{2}$ = 1.41421356..., so we get this."

$$\sin 45° = \frac{\sqrt{2}}{2} = \frac{1.41421356\cdots}{2} = 0.70710678\cdots$$

"I imagine that's more $\sin 45°$ than I'll ever need."

9.1.4 $\sin 60°$

"Try figuring out $\sin 60°$ then," I said.

Yuri scribbled in the notebook. "That would mean finding this y, right?"

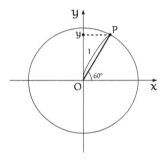

"Yep. Do you notice anything interesting about where the point is?"

Yuri took a long, hard stare at the graph. She scratched the tip of her nose and muttered, "No, not that," to herself, but she had that look I knew meant she wouldn't give up soon.

"There we go," she said, adding point A to the graph and connecting it and point P. "Yeah, that's gotta be it."

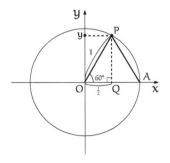

"Equilateral triangle, right?" Yuri said.

"Equilateral triangle indeed," I said. "Sides $\overline{OP}$ and $\overline{OA}$ here are radii of the same circle, so they have the same length. That means angles $\angle OPA$ and $\angle OAP$ have the same measure. We know that $\angle POA$ is 60°, and the sum of angles $\angle OPA$, $\angle OAP$, and $\angle POA$ has to be 180°, so that means they all have to be 60°."

"Like I said. So now we can drop this line straight down from vertex P to make triangle $\triangle POQ$. Side $\overline{OQ}$ is half as long as $\overline{OA}$, so its length is $\frac{1}{2}$. So the value of y must be, let's see...take the root of $1^2 - (\frac{1}{2})^2 \ldots$"

Yuri jotted some calculations in a corner of the page.

"Got it. $\overline{PQ} = \sin 60° = \frac{\sqrt{3}}{2}$."

"Good job," I said. "And $\sqrt{3}$ is approximately 1.7, so $\sin 60°$ should be something like 0.85."

Yuri nudged the calculator my way and raised her eyebrows.

"Fine, fine," I said, tapping out the calculation. "Here you go."

$$\sin 60° = \frac{\sqrt{3}}{2} = \frac{1.7320508\cdots}{2} = 0.8660254\cdots$$

"So what's next?"

"I guess $\sin 30°$. That's an easy one to do now, though."

"Easy how?"

"Look at the graph."

Yuri soon saw the trick.

"Oh, I get it. You just have to put the triangle on the y-axis and use $90° - 60° = 30°$. So $\sin 30°$ should be $\frac{1}{2}$, right?"

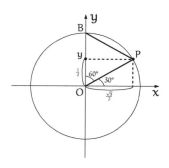

$$\sin 30° = \frac{1}{2}$$

"And there you have it," I said. "Values of $\sin \theta$ for $0°, 30°, 45°, 60°,$ and $90°$. Beyond that, you can just use symmetry to find the other major ones."

"Symmetry?"

"Left–right symmetry, over the y-axis. Like, $\sin 120°$ will have the same value as $\sin 60°$. Here, it's easier to see as a graph."

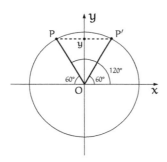

"Hmm... Ah, okay. Because P and P' will have the same y-coordinate. Gotcha. So I guess that works with all of these? Oh, and adding negative signs once you go past $180°$, I guess."

"Exactly," I said. "Here, let's summarize all of them at once."

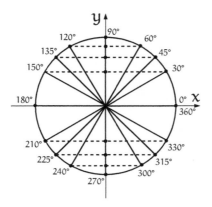

9.1.5 *The Sine Curve*

"Hey, you haven't told me about the sine curve yet," Yuri said.

"Why do you think we're figuring out all these values for sine?" I asked.

"You tell me."

"To plot out a sine curve. Look, let's show the relation between θ and $\sin \theta$ as a table."

θ	$0°$	$30°$	$45°$	$60°$
$\sin \theta$	$0.000 \cdots$	$0.500 \cdots$	$0.707 \cdots$	$0.866 \cdots$

θ	$90°$	$120°$	$135°$	$150°$
$\sin \theta$	$1.000 \cdots$	$0.866 \cdots$	$0.707 \cdots$	$0.500 \cdots$

"Yep, yep," Yuri said, nodding.

"Then after $180°$ we just add negative signs, yeah?"

θ	$180°$	$210°$	$225°$	$240°$
$\sin \theta$	$-0.000 \cdots$	$-0.500 \cdots$	$-0.707 \cdots$	$-0.866 \cdots$

θ	$270°$	$300°$	$315°$	$330°$
$\sin \theta$	$-1.000 \cdots$	$-0.866 \cdots$	$-0.707 \cdots$	$-0.500 \cdots$

"You betcha." Another nod.

"And when we get to $\sin 360°$ we start over at 0. So do you see the sine curve?"

"Not a one."

"Okay, let's plot these values out then."

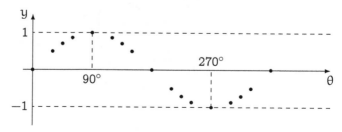

Plotting values of $(\theta, \sin \theta)$

"Whoa!" Yuri said, leaning forward to peer at the plot. "Things sure are starting to look pretty curvy here!"

"They'll be curvier if we connect the dots." I started to do so, but Yuri snatched the pencil from my hand.

"I've got this."

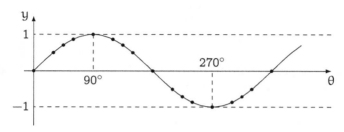

The sine curve

"There ya go," I said.

"The sine curve at last," Yuri said. But the satisfaction in her face soon faded.

"Something wrong?" I asked.

"I must be missing something. This is basically just the unit circle we've graphed, right? Why does it look so different?"

"You're right, you are missing something. Whenever you're looking at a graph, be sure to pay attention to the axes. When we graphed the unit circle we used x for the horizontal axis and y for the vertical axis. In other words, that graph showed us the relation between x and y. But the horizontal axis here is θ, so the sine curve is a relation between θ and y."

"Gimme the short-and-sweet version of all that."

"Sure, I'll write it down for you."

· The **unit circle** is a graph of how point P moves as a relationship between x and y.

· The **sine curve** is a graph of how point P moves as a relationship between θ and y.

"Much better," Yuri said. "Okay, I'm down with the sine curve."

"Good to hear."

"I'm starting to think it's over-rated, though. After all, with the unit circle graph you can see x, y, and θ, but with the sine curve you lose the x."

"I think they both have their uses."

Yuri took off her glasses and folded them up. She started speaking slowly, as if choosing her words.

"I noticed something else, too."

"About the sine curve?"

"No, about me. I think I tend to rush ahead too much. Like when I wanted to jump ahead to the cosine before I really understood the sine. I get . . . I don't know, hasty. Like, when I think I get something right away, I want to rush on to the next thing. When I don't get something right away, it seems like too much work so I ignore it. I need to slow down and work through things more, like you do."

"Well, you're right—there's no need to rush. After all it took hundreds, no *thousands* of years for some of the finest minds of each era to build up mathematics to what it is today. When everything's presented so nicely in a textbook as symbols and equations and proofs, it's easy to forget the amazing amount of work behind it all. So yeah, it's cool if you don't get everything right off the bat. It's better that way, even."

"Better?"

"Sure. Even when you're pretty sure you understand a chunk of math, it's still good to assume there's more there that you haven't seen yet."

"Better the love that burns low and long, than that which flares and fades, right?"

"Huh?"

"Never mind, I've clearly exceeded your comfort zone. More to the point, I think I'm ready for cosines now."

"Easy enough—just like the sine function is about the relationship between θ and y, the cosine function is the relationship between θ and x."

"That's it?"

"That's it, but—"

"—don't assume there isn't more there to learn about, right?"

"You're catching on."

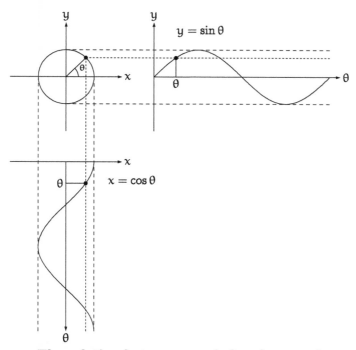

The relation between $y = \sin\theta$ and $x = \cos\theta$

9.2 $\frac{2}{3}\pi$ RADIANS

9.2.1 Radians

The next day Tetra and I had lunch up on the roof. I was in a great mood; the weather was gorgeous, and tomorrow would be the last day before the school year was over and I could enjoy spring break. The subject of Yuri had come up, and what we'd talked about the previous day.

"Yuck," Tetra said. "I suck at trig."

"Yeah? I thought you were pretty good at it."

"At bits and pieces, maybe, but I'm a long way from mastering it."

"I'm not sure there's *any* area of math I feel like I've mastered."

"Let me put it this way, then—I barely get it at all."

"Oh, come on."

"Seriously. Even radians are still something of a mystery to me."

"They take some getting used to, yeah."

"But they're just a way of measuring angles, right?" Tetra held up a chopstick and rotated it to measure off angles. "90° is $\frac{\pi}{2}$ radians, 180° is π radians, 360° is 2π radians, and so on. So I know they're something like normal degrees, but still... 360° is 2π radians? Why? I don't get where all that comes from."

"They're not so hard if you just think of them as the lengths of arcs measured in radii."

"See? You've already lost me."

"Nah, you can do it. Let's think about how many radians are in 360°. If the radius of a circle is r, then what's the length of an arc that goes around 360°? In other words, the full circumference of the circle."

"Well, the circumference of a circle should be $2\pi r$, right?"

"Sure. And $2\pi r$ is what multiple of r?"

"Um, if you divide $2\pi r$ by r I guess you just get 2π... And that's where the 2π radians comes from?"

"It is. So radians are just a way of measuring angles using arc lengths. But when a circle's radius changes, the angles stay the same while its circumference changes, so you have to think in terms of multiples of the radius."

"But why bother? Why not just use 360° degrees like normal?"

"Well, there's nothing special about 360, except that sometimes it's a convenient number because it has lots of factors. So there's nothing *wrong* with degrees, but sometimes it's also convenient to represent angles using something a little less arbitrary, like the ratio of a circle's circumference to its radius. Not that that's not somewhat arbitrary itself, I suppose."

"If you say so."

"But it kind of makes sense, doesn't it? An angle at the center of a circle determines some length along the circumference, and you can represent that as a multiple of the radius. That length for a 60° angle is $r \times \frac{\pi}{3}$, like this."

I pulled out a notebook and wrote the calculation.

$$2\pi r \times \frac{60°}{360°} = 2\pi r \times \frac{1}{6}$$
$$= r \times \frac{\pi}{3}$$

"So that means the length is $\frac{\pi}{3}$ times the radius. In other words, 60° is the same as $\frac{\pi}{3}$ radians."

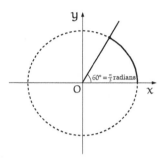

60° is the same as $\frac{\pi}{3}$ radians

"Somehow it seems less mysterious with the picture," Tetra said.

9.2.2 Teaching

Tetra finished her lunch and wrapped her bento box in a pink handkerchief. I shoved the wrapper from the bun I'd eaten into my pocket, then stood and stretched.

"I never thought it would happen," Tetra said, "but I've actually had friends coming to *me*, asking for help with math."

"Good for you! Teaching something can be the best way to learn it."

"Maybe I'm not quite there yet—half the time I get stuck explaining something, and they give up on me."

"That happens too," I said, nodding.

"It's funny how learning something and teaching it can be so similar in some ways, and completely different in others. Teachers must have it really tough. As many times as I've complained about how I'm being taught math, now I'm getting a better idea of how hard it can be. And to dozens of people at once, to boot!"

"It can't be easy," I said.

"*You* make it look easy, though. That's quite a talent you have."

"I'm not sure I could do it in front of a classroom, though. Teaching you is easy, because you pay attention and ask questions, and let me know what you're not getting. Without that, I'd never know if you understood me or not."

I trailed off into my own thoughts. As I got deeper and deeper into math, what I learned would no doubt become harder and harder to get across to others. I knew that if I ever made it to the cutting edge of mathematical knowledge there would be precious gems to dig out of rich mines, beautiful shells to collect along new beaches... But would I be able to show those treasures to others?

"Hello? You in there somewhere?"

I snapped back at the sound of Tetra's voice.

"Sorry, I spaced out for a minute there."

Tetra started plucking at the knot she'd tied in her handkerchief, arranging and smoothing it.

"I'm so glad I got to come to this school," she said.

"Well I'm glad you're here too."

"I'm also glad I wrote you that letter."

"Me too."

"And I...I, uh..."

The warning bell rang. Lunch break was over.

"And I...I..." Tetra bit her lip. "I'll see you later!"

She darted ahead of me, down the stairs and off to class.

9.3 $\frac{4}{3}\pi$ Radians

9.3.1   Classes Canceled

When I got back to my classroom, Miruka was standing guard at the doorway.

"Afternoon classes are canceled for us," she said.

"Wha—?"

She grabbed my arm and pulled me down the hall, out the exit, and all the way off the school grounds. I had to scramble to keep

up. We made our way down to the main road and crossed the intersection, heading for the train station. It felt weird to be out in town
at this unaccustomed hour.

A few minutes later, I found myself on a train, sitting next to
Miruka.

9.3.2 Remainders

The train seemed to move slowly through the balmy sunshine. I had
no idea where we were headed.

"So to state the obvious, you're the one that canceled our classes,
right?"

Apparently this was too obvious to warrant an answer.

"Where were you during lunch?" Miruka asked, wiping her glasses
with a cloth.

"Up on the roof."

"Hmph," Miruka said, putting her glasses back on and looking
straight at me.

"I was just having lunch with Tetra." The apologetic tone in my
voice surprised me.

"She's a sweet girl."

"We talked about radians."

"She's a sweet girl," Miruka repeated.

"Y'know, how $360° = 2\pi$ radians and stuff."

"She's a sweet girl."

"Uh, yeah..." I nodded.

"You told her about $\theta \bmod 2\pi$?"

"I...er, what?"

"About how everything repeats."

"I have no idea—"

Miruka made a familiar hand-it-over gesture, so I pulled a notebook and pencil out of my bag. She took them and wrote a single
expression.

$$\theta \bmod 2\pi$$

I looked at it, trying to focus enough to think—the oddness of
the situation seemed to be shutting down the math center of my
brain.

a mod m. *The remainder when you divide a by m, a.k.a. the remainder modulo m. Like, 17 mod 3 = 2, because when you divide 17 by 3, you get a remainder of 2. Normally a and m should both be integers, but Miruka wrote θ mod 2π. So this is the remainder when you divide θ by 2π? What's a real modulo another real? Does this make sense?*

I glanced at Miruka. She was looking out the window, pretending to watch the scenery pass by, but I knew her attention was fully on me.

Using θ here implies that this is an angle. So an angle modulo 2π would be...?

"Oh, I get it," I said. "You're spinning the point around and around the circle. You're talking about how when the point has rotated by θ radians, it ends up at the same place as if it had moved θ mod 2π radians."

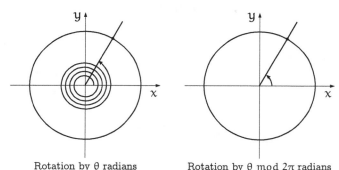

Rotation by θ radians   Rotation by θ mod 2π radians

Miruka looked back to me and nodded.

"So for two real numbers a and b, you get this relationship."

$$a \bmod 2\pi = b \bmod 2\pi$$

"In other words," she continued, "a and b are congruent, modulo 2π. Easier to write it this way, though."

$$a \equiv b \pmod{2\pi}$$

"This relationship satisfies the reflexive, symmetric, and transitive properties, so it's an equivalence relation. You can use this relationship to partition the set $\mathbb{R}$ of all reals."

I nodded and swallowed, unsure how else to respond given the weird circumstances.

"When we look at an angle θ, what we're actually seeing is an element of a quotient set."

$$\{2\pi \times n + \theta \mid n \text{ is an integer}\}$$

"A representative of infinitely many possible angles, superimposed on one another."

"Yeah, that's cool," I said. "Equivalence relations and quotient sets pop up all over the place, I guess."

Miruka stood from her seat.

"What's wrong?" I asked.

"This is our stop."

9.3.3 The Lighthouse

When we stepped off the train, the air was filled with the salty tang of the ocean.

"This way," Miruka said.

She led me out of the station and down a series of twisting, narrow roads. She didn't look back until we finally broke out again into the open. The glare of a beach stung my eyes.

"There," Miruka said. I looked and saw a lighthouse, almost painfully white against blue sky and sea.

We reached the lighthouse and found the door open. Miruka started up the spiral staircase within, leaving me with nothing to do but follow. Around and around we went, me hurrying to keep up.

At the top was another door. When we passed through, we were outside again, overlooking the ocean below. The thin line of the horizon stretched across my field of vision. The sun twinkled and flashed off of countless waves. The view took my breath away.

It was too early in the season for swimming, so the beach below was empty. A gentle breeze was blowing into shore, carrying a strong smell of brine.

"I've received an invitation to study abroad," Miruka said.

"You've—*what*? Where? From who?"

"A mathematician, Dr. Narabikura. My aunt. She wants me to come study at the university where she does research, once I'm done with high school. It's in the U.S."

A seagull took off from a rock below and headed out over the water. I watched it dip and weave until it was out of sight.

"Have you made a decision?"

"Yes."

"You're going, aren't you."

Miruka nodded. I felt something like ice in my chest.

"It'll be tough, but I think it's the best place for me."

I was reeling in shock, but I couldn't say exactly why. Had I really thought we'd end up at the same college? I still had no idea what I was going to do after high school. But this... She was forcing me to face a fact I'd been trying to ignore—once school was done with, we'd end up going our separate ways.

"I see."

Miruka turned to me. The wind blew her long hair into rippling waves.

"You see what?" she asked.

"You've been keeping this to yourself for a while now, haven't you. I had no idea. I...I thought you trusted me more than that."

I felt my words turn bitter toward the end.

"I didn't say anything because I hadn't decided yet," she said.

"And now you've decided you don't want anything to do with me? Is that what you're saying?"

"Of course not. I just wanted you to know."

"Well now I know."

9.3.4   On the Beach

We made our way back down the stairs and walked side-by-side along the beach in silence. I watched the waves trying to climb up onto the sand, only to recede again. Up and back, up and back, over and over again. Sometimes depositing clumps of seaweed or unidentifiable debris, but mostly with nothing to show for their efforts.

Of course Miruka would end up studying overseas. With her talent she could go to any school she wanted to, in the U.S. or any country. They'd be lucky to have her.

I wish I could say the same for myself...

A wave of disgust washed over me. Someone incredibly important to me was about to accomplish something wonderful, and all I could do was feel sorry for myself.

I'm such a child. I'll always be such a child. How frustrating. How pathetic...

Something struck my left cheek, and a moment later I felt a flash of pain. It took me a few seconds to realize what had happened.

"You idiot!" Miruka shouted.

"What—?"

She raised her hand again so I stepped back out of range. I adjusted my glasses, which had gone askew when she'd slapped me.

"I said you're an idiot, idiot. Don't try to tell me that you weren't just sulking about how worthless and pathetic you are. What's that gonna accomplish, huh? Making yourself miserable isn't going to change anything."

"I just—"

"Don't even start. You're smart, so take a look around you and use your head. You're surrounded by people who care about you. Tetra, Yuri, Ay-Ay, your mother... You affect everyone around you when you mope around like that. So cut it out!"

"But I—"

"You nothing. You do this all the time. You revel in meaningless depression, and I'm sick of it."

"No, seriously. Something's not right. I'm not getting anywhere, I'm not moving toward anything! I'm just spinning in circles like a little kid."

Miruka stopped and looked at me for a minute. When she spoke again, her voice was softer.

"You aren't seeing all the dimensions."

"Huh?"

"You just see a point going around and around, because you're looking at things from the wrong angle. It's preventing you from seeing the spiral." She looked down at the sand beneath our feet. "Cheer up. Good things are all around you."

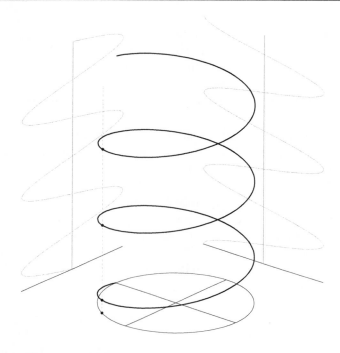

9.3.5 Salving the Wound

We stood there for a minute, Miruka staring at the sand, and I at her. My cheek burned and started to throb. She had helped me work something out of my system, though. She was right, of course—my moping wouldn't help anything. It was a habit I needed to kick.

So she would be gone once we were out of school. That was just something I had to deal with. Starting now.

"Hey, Miruka."

She looked up at me.

"Forgive me? For being such a jerk sometimes?"

"Hmph," was all she said, but she didn't break eye contact.

"I'll cut out the moping," I said. "I'll do my best, at least."

"You're bleeding," she said, pointing at my cheek.

"Yeah?" I touched my cheek, and my fingertips came away red.

"I must have scratched you."

"I'm okay."

"Here, let me."

Miruka grabbed my head with both hands and pulled me toward her. When we were nearly nose-to-nose she turned my face and ran her tongue up my cheek.

"You taste like the sea," she said, giving me the warmest smile I'd seen from her in a long time.

If you want to be a mathematician, you must realize you will be working mostly for the future.

ALFRÉD RÉNYI
Dialogues on Mathematics

CHAPTER 10

Gödel's Incompleteness Theorems

> All truths are easy to understand once they are
> discovered; the point is to discover them.
>
> GALILEO GALILEI

10.1 AT THE NARABIKURA LIBRARY

10.1.1 Entrance

Soon after spring break began, Yuri, Tetra, and I gathered at the
Narabikura Library, a three-story building that stood high on a hill.
After making the arduous climb from the train station, we paused
to take a breath. I looked up at the odd dome that sat atop the
building.

"Is this it?" Tetra asked.

"Better be," said Yuri. "I'm exhausted."

"Rest easy," I said, pointing to the carved sign above the doorway.

Heading inside, we entered a large atrium that looked more like a hotel lobby than a library. Each floor wrapped around the central open space, forming stacked circles of glass-fronted bookshelves. Comfortable-looking sofas were scattered about, some with readers deep in their studies. It had that wonderful smell of countless old books that you could only experience at a library.

"I wonder where we're supposed to go?" Tetra said, looking about.

"Miruka said she'd let us know," I said. "Haven't heard from her yet, though."

We asked at the front desk, where a sharply-dressed young man directed us to the 'Chlorine' room on the ground floor.

"Did you see that guy?" Yuri said to Tetra as we walked down a hallway. "What a hottie."

"I wonder if he's a librarian?" Tetra said.

"I wonder if he's *single*?"

I rolled my eyes but knew well enough to keep my mouth shut.

"Ah, here it is," I said, stopping in front of a door marked 'Cl.' I knocked on the door and opened it, and found Miruka sitting inside.

10.1.2 Chlorine

"There you are," Miruka said.

"Here we are," I said. "But where exactly is *here*?"

I sat in a nearby chair and surveyed the room. There was an oval table in the middle, surrounded by several high-backed chairs. A giant white board covered one full wall, and in the corner was a small table with a complicated-looking telephone. A wide window opened to a garden-like green space. It felt more like an office than a library study room.

"The Narabikura Library, of course," Miruka said. "It's a private library, and a good one. Lots of technical books, nice meeting rooms. It even hosts international conferences and special research sessions from time to time. I attend all the mathematics ones I can."

So that's what Miruka does in her spare time, I thought.

"So what are we here for?" Yuri asked.

"To do math on math," Miruka replied.

"Meaning?"

"I want to think about Gödel's incompleteness theorems, the starting point of modern logic."

"Why do it here?"

"Because classrooms aren't as comfortable, and we're going to be here for a while. And apparently our school is off-limits to you."

"You're so sweet!" Yuri squealed. "Speaking of which, I brought you something."

Yuri dug a box of chocolates out of her bag and handed it to Miruka.

"Thanks," Miruka said. "I'll need the fuel."

Miruka opened the box and popped a chocolate in her mouth.

"This is going to be something of a marathon," she said, "so let's begin with a road map of where we're going."

Tetra opened a notebook to a fresh page. "Good idea!"

"We'll start with something called Hilbert's program, an effort by the mathematician David Hilbert to develop a solid foundation for mathematics. Then we'll run through Gödel's incompleteness theorems, which showed that Hilbert's program as it was originally proposed was already dead in the water. We'll talk about what the theorems are all about first, then walk through them. That's going to take a while, so we'll have a lunch break in there somewhere, but in the meantime feel free to snack."

Miruka reached under the table and pulled out a bulging plastic bag. She emptied it onto the table, creating a small mountain of junk food.

"Finally, I want to talk about what these theorems really *mean*. There's a lot of misinformation out there, about how these theorems destroyed Hilbert's program, how they say something about the limits of mathematics... But what they really do is provide the basis for modern logic. There's a lot more that's positive about these theorems than negative."

Miruka paused and looked at each of us in turn.

"Everyone ready to get started?" she asked.

Tetra, Yuri, and I nodded in unison.

"Very well, then. Let's begin."

10.2 HILBERT'S PROGRAM

10.2.1 Hilbert

"David Hilbert was a leading mathematician from the late nineteenth to the early twentieth century," Miruka began. "He's the one who proposed Hilbert's program to provide a solid foundation for mathematics. The program was to have three major phases."

Miruka went to the whiteboard and began writing.

Introducing formal systems
  Representing mathematics as a formal system

Proving consistency
  Proving that there are no contradictions in the formal system representation of mathematics

Proving completeness
  Proving that the formal system representation of mathematics is 'complete'

"I think I'm going to need some more details," Yuri said.

"Don't worry, I'll explain each of these in order. So, introducing formal systems . . . "

Miruka sat in the chair nearest the board.

"Mathematics is a huge subject, one that spans many areas. So the first step in creating a foundation for it is defining exactly what it *is*. Hilbert decided to attempt that by viewing mathematics as a formal system."

"Like we did before?" Tetra asked.

"Just like that. Setting up propositional formulas as series of symbols, deciding which propositional formulas should be axioms . . . then what?"

"Deciding on inference rules," I said. "You use those to turn formulas into other formulas."

"Oh, I remember this!" Tetra said. "Then you can start from the axioms and apply inference rules to string together formulas and create proofs!"

"*Formal* proofs," Miruka said. "And what did we call the last formula at the end of a formal proof?"

"Um ... a theorem!"

"Good. So if you have formal proofs in a formal system, and in some sense those represent proofs in mathematics, then you can claim that your formal system has captured some aspect of mathematics. And if you can represent mathematics as a formal system, then the research you do on your formal system should apply to mathematics in general."

Miruka reached across the table and picked out another chocolate.

"Okay, moving on to proving consistency," she said. "Hilbert figured that once we had a formal system that represented mathematics, next we'd have to show that the system didn't contain any contradictions. What's a contradiction in this case, Tetra?"

"When you can prove both A and $\neg A$ for some statement A, right?"

"Right. In a formal system that contains contradictions, you can find a way to prove *any* propositional formula, so it isn't a very useful system. But if you could show that a formal system representation of mathematics was consistent, then you could rest assured there would be no formal proofs of both A and $\neg A$. Of course that's quite a claim to make, so you'd want to be sure there's absolutely no question that your proof was valid. With that in mind, Hilbert wanted to use a formal system devoid of all meaning and give a clear proof of that system's consistency. Good so far?"

Tetra and I nodded. Yuri looked a bit nervous, but she brightened when Miruka gave her a reassuring wink.

"On to proving completeness, then. Hilbert knew that showing there were no contradictions in his system wasn't enough. He also had to show that his system was complete, meaning that for any formula A in the formal system, you could give a proof of A or of $\neg A$. So once mathematics was represented as a formal system, every valid mathematical statement could be shown to be at least one of true or false."

Miruka sat back and crossed her arms.

"So that was Hilbert's plan—represent mathematics as a formal system, and show that for any formula in the system you couldn't prove that the formula was both true and false, but you could prove it was one or the other. The result would be a mathematics in which there was no darkness that couldn't be illuminated by the light of a formal proof. It would be the basis for all mathematics that followed, forever more."

Hilbert's Program

Introduce formal systems
Represent mathematics as a *formal system*, allowing it to be written as strings of symbols.

Prove consistency
Show that the formal system is *consistent*, in other words that for all propositional formulas A in the system you could not give formal proofs of both A and $\neg$A.

Prove completeness
Show that the formal system is *complete*, in other words that for all propositional formulas A in the system you could give a formal proof of at least one of A or $\neg$A.

10.2.2   Quizzes

"Um, Miruka?" Yuri said. "What's the difference between proofs and formal proofs?"

"Hmph. *Somebody* was supposed to catch you up on all this." Miruka shot me an icy stare.

"C'mon Yuri, we talked about this just the other day!" I said.

"I'll admit the words do sound familiar."

"A quick refresher then," Miruka said. "Like we just said, in a formal system you use strings of symbols to create propositional formulas. You set aside some formulas to act as axioms, and you decide on some inference rules that use formulas to create new formulas. You're good with that so far?"

Yuri gave a firm nod.

"Okay. So a formal proof is a finite sequence of propositional formulas $a_1, a_2, a_3, \ldots, a_n$ that fulfills these conditions."

· a_1 is an axiom.

- a_2 is either an axiom or a formula derived from a_1 using one of the inference rules.

- a_3 is either an axiom or a formula derived from a_1 and/or a_2 using one of the inference rules.

- ...

- a_n is either an axiom or a formula derived from one or more of a_1 through a_{n-1} using one of the inference rules.

"If that checks out for your sequence of formulas then it's a formal proof, and that last formula a_n is a theorem in the formal system. So in the end, a formal proof is just a specific kind of a sequence of formulas, and one that's limited to the formal system you're dealing with. Consider it a concept that exists in a 'formal world,' if you like."

I nodded, secretly glad to hear that I had all this down myself.

"So what's a 'normal' proof?" Yuri asked.

"The kind of proof you usually see when you're doing math. One that isn't tied to a formal system. One that exists in the 'world of meaning.' Watch out, though—when the context is understood, it isn't uncommon to abbreviate 'formal proof' to just 'proof.'"

"Okay, got it."

"Let's have a quick quiz to make sure, then."

Yuri shot straight up in her seat.

"Um, okay ... "

"Answer me this: in a formal system, can you consider an axiom to also be a theorem?"

"I, uh ... don't know."

"What do you think, Tetra?"

"I think you can," she said. "You said a theorem is the last line in a formal proof, right? But there's nothing that says how long a proof has to be. So I could just say that some axiom a is itself a really short proof. Then a is the last line in that proof, which would mean it's a theorem. So doesn't that mean that all axioms are also theorems?"

Miruka nodded. "Well done."

"Makes sense," Yuri said.

"Another quiz, then," Miruka said. "Say you have some complete formal system X and a statement a that is *not* a theorem in that system. If you add that statement a to X to create a new formal system Y, then Y will contain a contradiction. Why?"

We all sat silently for a time.

"Uh...remind me what a statement is?" Tetra said.

"A propositional formula that doesn't contain any free variables," Miruka said.

More silence.

"Back to definitions, then," Miruka said. "What does it mean for a formal system to be complete?"

"That for any statement a, you can prove at least one of A or $\neg A$," I said.

"And what's the condition for a statement a to not be a theorem?"

"That there's no way to give a formal proof of it," Tetra said.

"And what makes a formal system contradictory?"

"It has a formula A where you can prove both A and $\neg A$," Yuri said.

"That should be all the hints you need," Miruka said. "So why does the formal system Y contain a contradiction?"

A third round of silence. I tried my best to mentally engage. *If you can't give a formal proof of a statement a, that would mean...*

"I got it!" Yuri shouted. Sunlight gleamed off of her chestnut ponytail as she bounced up and down in her seat.

"Let's hear it," Miruka said.

"Okay, so X is complete, which means you can prove at least one of a or $\neg a$." Yuri was speaking even faster than usual. "But if a isn't a theorem, that means you can't prove it. That means you *should* be able to prove $\neg a$. But you've added a as an axiom in Y, right? So that means you can prove both a and $\neg a$ in Y! A contradiction! Right? Am I right?"

Yuri looked expectantly at Miruka.

"Well done," Miruka said.

"Yesss!" Yuri gave a fist pump.

Score one for Yuri. That girl's a demon when it comes to logic problems.

Miruka held her arms out in a circle, as if holding a huge beach ball.

"So if you add even a single statement that can't be proven as an axiom in a complete formal system, you end up with a contradiction and the whole thing collapses."

Miruka's invisible beach ball deflated as she tightened her arms around it.

"One last quiz," she said. "If a formal system is contradictory, then it ends up you can prove *all* formulas in that system. I'm not going to prove that just now, but take my word on it."

We all nodded our assent.

"Okay, given that, *all contradictory formal systems are complete*. Why?"

"Oh, sure, that makes sense," I said.

"Huh? Contradictory, yet complete? How's that work?" Tetra asked.

"Careful now," I said. "Don't let the dictionary definitions of 'complete' and 'contradictory' trip you up. Miruka said that if a formal system is contradictory, then you can prove all formulas in it, right? But statements are just a specific kind of formula, so you should be able to give a formal proof of all statements. That makes the formal system complete. After all, being a complete formal system means that no matter what statement A you choose, you'll be able to prove at least one of A or $\neg A$. That's straight from the definition. But if you can prove *both* A and $\neg A$, that means you can definitely prove *at least one of* A or $\neg A$. So the formal system is complete."

Miruka smiled. "That'll do. I know it sounds weird to say that a contradictory system is complete if you're thinking in terms of everyday usage of those words, but if you stick to mathematical definitions it makes perfect sense."

Tetra nodded, mumbling aloud as she wrote in her notebook. "All contradictory systems are complete..."

"A word of warning," Miruka said, a stern look on her face. "We're talking math here, not philosophy or ethics. Don't try to pull this 'contradictory systems are complete' stuff into areas where it doesn't belong."

"I wouldn't dream of it," I said.

"Okay then, on to Gödel."

10.3   GÖDEL'S INCOMPLETENESS THEOREMS

10.3.1   Gödel

"Gödel published proofs of his incompleteness theorems in 1931, when he was 25 years old," Miruka said. "The full title of the paper they appeared in is *On Formally Undecidable Propositions in* Principia Mathematica *and Related Systems I*. It was originally written in German, but I have a translation here. Let me read part of the introduction."

Miruka pulled a thick bundle of papers out of her bag and began reading from the top page.

> The development of mathematics towards greater exactness has, as is well-known, led to formalization of large areas of it such that you can carry out proofs by following a few mechanical rules. The most comprehensive current formal systems are the system of *Principia Mathematica* (PM) on the one hand, the Zermelo–Fraenkelian axiom-system of set theory on the other hand. These two systems are so far developed that you can formalize in them all proof methods that are currently in use in mathematics, i.e., you can reduce these proof methods to a few axioms and deduction rules. Therefore, the conclusion seems plausible that these deduction rules are sufficient to decide all mathematical questions expressible in those systems. We will show that this is not true... [1]

---

[1] Translation from the original German by Martin Hirzel. Used with permission.

Miruka put the paper down and scanned our faces as she talked.

"Gödel gave proofs for many theorems in this paper, but there are two in particular that we call his 'incompleteness theorems' today. We refer to them as the 'first incompleteness theorem' and the 'second incompleteness theorem.' "

Miruka went to the whiteboard and began writing.

Gödel's first incompleteness theorem

In any formal system meeting certain conditions, there exists some statement A for which the following two conditions hold:

- There exists no formal proof of A in the system.

- There exists no formal proof of ¬A in the system.

Gödel's second incompleteness theorem

In any formal system meeting certain conditions, there exists no formal proof for a statement that asserts the system's consistency.

"These two theorems were quite a blow to Hilbert's program," Miruka said. "The whole thing would of course be a no-go if there were no way to give formal proofs of either completeness or self-consistency."

"But, 'certain conditions'?" I said, raising an eyebrow.

Miruka smiled and started to speak, but was interrupted by Tetra's raised hand.

10.3.2 Discussion

"Question!" Tetra said. "Wouldn't this second theorem imply that all of mathematics contains contradictions? That there was no way to show it was consistent?"

"*No*," Miruka said, a stony look in her eyes. "You're slipping back into sloppy language. Saying things like 'all of mathematics contains contradictions' and 'mathematics isn't consistent' is too vague to carry useful meaning. Read the second theorem one more time."

> In any formal system meeting certain conditions, there exists no formal proof for a statement that asserts the system's consistency.

"There's nothing in there about mathematics itself. This is strictly a theorem about formal systems meeting certain conditions. Not only that, but it only talks about proofs for self-consistency. So it's saying that when you're working within a formal system that meets certain conditions, you can't prove the consistency of that formal system itself. But there *are* cases where you can work from the outside, proving the consistency of a formal system from within the context of another."

"So it's like saying we can find inconsistencies in others, but not in ourselves?" Tetra asked.

"A very rough way of putting it, but sure, something like that. In any case, mathematics continues to purr happily along, second incompleteness theorem or no. Want to prove the consistency of a system? No problem—just use a slightly more powerful system to do so. It isn't even that hard a thing to do, in many cases. Gödel's incompleteness theorems sound extreme when you strip away all the mathematical conditions on them, and the word 'incomplete' has all kinds of emotional baggage when you take it out of the realm of mathematics. But by now you guys should know better than to apply mathematical assertions to non-mathematical arguments."

"But, Miruka," Yuri said, "I read somewhere that this was all about limits to human reasoning, or something like that."

"No, no," Miruka said, casting a soft look at Yuri. "Gödel's theorems are strictly *mathematical* theorems, and math has nothing to say about squishy topics like 'the limits of reason.'"

"Okay, if you say so."

"Don't take my word on it, think about it yourself. There's no real number solution to $x^2 = -1$, but that doesn't mean that equation somehow lies 'beyond the limits of reason.' It just shows us something about the nature of equations. The incompleteness theorems are the same—they show us something about the nature of formal systems of a certain type. Not that they didn't have a huge impact on the world of mathematics. They did. But it wasn't a negative thing that somehow diminished mathematics. It was a positive thing, one that made mathematics stronger."

"I have a question too," I said. "What are these 'certain conditions' that keep popping up?"

"Consistency, inclusion of Peano arithmetic, and recursion," Miruka said. "In other words, the system can't have any contradictions, it must include the natural numbers, and there has to be a way of mechanically determining if a formal proof is formed by a proper sequence of formulas. Gödel's paper used something called 'ω-consistency,' a stronger form of consistency than usual, but later Rosser showed that you could weaken that condition to simple consistency."

10.3.3   Outline of the Proof

"Let's take a look at an outline of Gödel's proof," Miruka said. "It's in five parts, which I'm going to call Spring, Summer, Fall, Winter, and the New Spring."

Spring: Formal system P
  Decide on symbols, axioms, and inference rules for a formal system P.

Summer: Gödel numbers
  Assign numbers to symbols and strings in P.

Fall: Primitive recursion
  Define a primitive recursive predicate, and introduce its representation theorem.

Winter: The long, hard journey to provability
  Define everything from arithmetic predicates to provability predicates.

The New Spring: Undecidable statements
  Construct an undecidable statement, one for which neither A nor $\neg$A can be proven.

10.4   Spring: Formal System P

10.4.1   Elementary symbols

"We start off in Spring," Miruka said. "Here we construct our formal system P. The P used in *Principia Mathematica*, which Gödel was specifically addressing, included the Peano axioms, along with a few others added in. It has symbols for addition, subtraction, exponents, and size relations."

"Size relations?" Tetra said.

"Inequalities. Greater than and less than," I said.

"Oh, of course."

"The end game here is to show that P contains an undecidable statement," Miruka continued, "but keep in mind that P is just one of infinitely many formal systems that the incompleteness theorems apply to. Also, when I use the word 'number' in this discussion I'm referring to an integer 0 or greater. Good?"

We all nodded.

"Okay. First off we need some elementary, or atomic, symbols. Two kinds actually: one for constants and one for variables."

"Isn't that cheating?" Tetra asked. "I thought we aren't supposed to think about what the symbols mean."

"You don't have to, and the math doesn't require it, but it will make everything easier to follow that way," Miruka said.

"Well I'm all for that!"

"Constants first, then."

Miruka went to the far left side of the board and started making a list.

▷ **C-1**

    0 ('zero') is a constant.

▷ **C-2**

    s ('successor') is a constant.

▷ **C-3**

    ¬ ('not') is a constant.

▷ **C-4**

    ∨ ('or') is a constant.

▷ **C-5**

    ∀ ('all') is a constant.

▷ **C-6**

    (('left parenthesis') is a constant.

▷ **C-7**

) ('right parenthesis') is a constant.

"And now, some variables," Miruka said, starting a new list below the first. "These come in different types."

▷ **Type-1 variables**

    are variables in the form x_1, y_1, z_1, ... and are used for numbers.

▷ **Type-2 variables**

    are variables in the form x_2, y_2, z_2, ... and are used for sets of numbers.

▷ **Type-3 variables**

    are variables in the form x_3, y_3, z_3, ... and are used for sets of sets of numbers.

"We can also create 'type-n' variables in a similar fashion, as needed. Also, there are only twenty-six letters in the alphabet, but we'll assume we can use any countable collection of characters in their place."

10.4.2 Numerals and Individuals

"Numerals are next," Miruka said. "Numerals are what we use to indicate numbers in our formal system P."

· The number 0 is represented by the numeral 0.

· The number 1 is represented by the numeral s0.

· The number 2 is represented by the numeral ss0.

· The number 3 is represented by the numeral sss0

· . . . .

· The number n is represented by the numeral $\underbrace{ss \cdots s}_{n \text{ of these}} 0$.

▷ **Numerals**
    are sequences of the form 0, s0, ss0, sss0, ...

.

"So these 's' characters perform the same role as primes when we worked with the Peano axioms, right?" Tetra asked.

"Exactly," Miruka said. "Onward we go, to defining individuals."

▷ **Type-1 individuals**
    are sequences of the form 0, s0, ss0, sss0, ..., or x, sx, ssx, sssx, ..., where x is a type-1 variable.

▷ **Type-2 individuals**
    are type-2 variables.

▷ **Type-3 individuals**
    are type-3 variables.

"I'm not sure I get this one," Tetra said.

"In other words, a type-1 individual will be something like sss0 or sssx$_1$," Miruka said. "Type-2 and type-3 individuals will just be variables like we said before, a set of numbers like x$_2$, or a set of a set of numbers like x$_3$. Oh, and here too we can go as far as we want, calling a type-n variable a type-n individual."

10.4.3   Elementary Formulas

"Time to define elementary formulas," Miruka said.

▷ **Elementary formulas**
    are sequences of symbols of the form $a(b)$, where a is a type-$\{n + 1\}$ symbol, and b is a type-n symbol.

"So for example x$_2$(0) and y$_2$(ssx$_1$) and z$_3$(x$_2$) are elementary formulas."

"So they come in the form set(element)?" I asked.

"If you want to look at it that way," Miruka said.

"Which means I can expect $x_2(x_1)$ to mean $x_1 \in x_2$?"

"You can, if you keep in mind that x_1 will be a number, and x_2 will be a set of numbers."

"Will do," Tetra said.

"On to formulas, then."

▷ **F-1**

Elementary formulas are formulas.

▷ **F-2**

If a is a formula, then $\neg(a)$ (the *negation* of a) is a formula.

▷ **F-3**

If a and b are formulas, then $(a) \vee (b)$ (the *disjunction* of a and b) is a formula.

▷ **F-4**

If a is a formula and x is a variable, then $\forall x(a)$ (the *generalization* of a) is a formula.

"Hey, I've seen this before!" Tetra said. "It's just like when we defined propositional formulas in formal systems!"

Miruka nodded and proceeded on. "Let's define some abbreviations."

▷ **Abbr-1**

We define $(a) \Rightarrow (b)$ to mean $(\neg(a)) \vee (b)$.

▷ **Abbr-2**

We define $(a) \wedge (b)$ to mean $\neg((\neg(a)) \vee (\neg(b)))$.

▷ **Abbr-3**

We define $(a) \Leftrightarrow (b)$ to mean $((a) \Rightarrow (b)) \wedge ((b) \Rightarrow (a))$.

▷ **Abbr-4**

We define $\exists x(a)$ to mean $\neg(\forall x(\neg(a)))$.

"Abbreviations? In math?" Yuri said.

"Just to make things simpler," I said. "It's a lot easier to write $(a) \Rightarrow (b)$ instead of $(\neg(a)) \vee (b)$, right?"

"And easier is better. Okay, I'm good."

"Just one more then, to make things even easier on us," Miruka said.

▷ **Parentheses**

will be abbreviated when doing so increases legibility.

10.4.4 Axioms

"We're also going to want to include the Peano axioms in our formal system P," Miruka said.

▷ **Axiom I-1**

$$\neg(sx_1 = 0)$$

▷ **Axiom I-2**

$$(sx_1 = sy_1) \Rightarrow (x_1 = y_1)$$

▷ **Axiom I-3**

$$x_2(0) \wedge \forall x_1(x_2(x_1) \Rightarrow x_2(sx_1)) \Rightarrow \forall x_1(x_2(x_1))$$

"I seem to remember there being five of those," Tetra said.[2]

"We've already included the first two from the standard list, when we defined variable types," Miruka said.

"We haven't defined the equals sign, though," Yuri said.

"Gödel did that by reference to the *Principia Mathematica*, which defined it by saying that $x_1 = y_1$ means $\forall u_2(u_2(x_1) \Rightarrow u_2(y_1))$. In other words, 'for any set u, if x_1 is a member of the set, then y_1 is too.'"

"Huh?"

"It says that if there's no set that only one of x_1 or y_1 belongs to, then they're equal. And that works all the way up to type n variables."

"Okay, I'm seeing that now."

"Good. Next we add the propositional axiom schemata. For all formulas p, q, and r, if they fit into one of these four patterns then they're axioms."

---

[2](cf. p. 25)

▷ **Axiom II-1**

$$p \lor p \Rightarrow p$$

▷ **Axiom II-2**

$$p \Rightarrow p \lor q$$

▷ **Axiom II-3**

$$p \lor q \Rightarrow q \lor p$$

▷ **Axiom II-4**

$$(p \Rightarrow q) \Rightarrow (r \lor p \Rightarrow r \lor q)$$

"And along with that we need schemata for some predicate logic axioms."

▷ **Axiom III-1**

$$\forall v(a) \Rightarrow subst(a, v, c),$$

where

- $subst(a, v, c)$ is a formula in which each instance of the free variable v in a is replaced with c,

- c is an individual of the same type as v, and

- within a, in the range over which v is free, there are no bound variables in c.

"Um, I think this one's going to require some explanation," I said.

"Even better, an example," Miruka said. "Remember, we've substituted all the v's in a with c."

- Let a be the formula $\neg(x_2(x_1))$.

- v is the type-1 variable x_1.

- c is the type-1 individual (numeral) $s0$.

- Then $subst(a, v, c)$ is the formula $\neg(x_2(s0))$.

"So it's sort of like doing a search-and-replace over a formula?"

"With some restrictions, sure," Miruka said. "Another predicate logic axiom schema."

▷ **Axiom III-2**

$\forall v(b \lor a) \Rightarrow b \lor \forall v(a),$

> where v is some variable, and no free v appears in b.

"No v's are allowed in b so that the $\forall v$ doesn't affect things, I guess?"

Miruka nodded. "Here's a fourth schema, for the inclusion axiom."

▷ **Axiom IV**

$\exists u(\forall v(u(v) \Leftrightarrow a)),$

> where u is a type-$\{n + 1\}$ variable not free in a,
> and v is a type-n variable.

"Inclusion?" I said.

"Intensional definitions of sets, in other words."

"Come again?"

"This lets us define a set u using some formula a."

"Ah. Okay."

"One final axiom, so that we can bring extensional definitions to our formal system P too."

▷ **Axiom V**

$\forall x_1(x_2(x_1) \Leftrightarrow y_2(x_1)) \Rightarrow (x_2 = y_2)$

"We'll take this formula and any other that's 'type-lifted' from this formula as an axiom. A type-lift is when you increase all symbol types by the same amount. Something like this."

· $\forall x_1(x_2(x_1) \Leftrightarrow y_2(x_1)) \Rightarrow (x_2 = y_2)$

· $\forall x_2(x_3(x_2) \Leftrightarrow y_3(x_2)) \Rightarrow (x_3 = y_3)$

· $\forall x_3(x_4(x_3) \Leftrightarrow y_4(x_3)) \Rightarrow (x_4 = y_4)$

· ...

"Erm, what's the point of this one?" I asked.

"Say that for every x_1 you choose, x_1 either belongs both to set x_2 and to set y_2, or it doesn't belong to either one. Then you could say that sets x_2 and y_2 are equal, right?"

"Ah, okay. And you said this had something to with extensional definitions?"

"Of sets, right. Because it's saying that a set is determined by its elements."

10.4.5 Inference Rules

"The last thing our formal system P needs to be complete is some inference rules," Miruka said.

▷ **IR-1**

a and a $\Rightarrow$ b give b.

Here, b is called the *immediate consequence* of a and a $\Rightarrow$ b.

"That's *modus ponens*, right?"

"It is. In this formal system we also have one more inference rule."

▷ **IR-2**

a gives $\forall v(a)$.

Here v is an arbitrary variable, and $\forall v(a)$ is called the *immediate consequence* of a.

"I'm not sure what's going on here," Tetra said.

"It says that if you can derive a with no conditions, then you can add a 'for all' condition to it."

Miruka skimmed the board for a final check.

"We're done with Spring," she said. "We've defined our formal system P. The next season is Summer, when we'll take on Gödel numbers."

She put down her marker.

"But before that, lunch."

10.5 LUNCH BREAK

10.5.1 Metamathematics

We followed Miruka up to the third floor and into a room labeled 'Oxygen,' which was something like a café. We took an outside table on the terrace, which afforded a view of the ocean in one direction, and a forest in the other. The skies were clear, and the sunlight was soft and warm.

I ordered curry, Yuri spaghetti, and Tetra a sandwich. Miruka just had a slice of chocolate cake.

"All this work with formal systems sure has changed my image of logic," I said.

"How's that?" Miruka said.

"Not long ago if I heard the word 'logic' I'd just think of things like syllogisms and De Morgan's laws, but clearly there's a lot more to it. I never thought it would be a field that let you study mathematics itself."

"Well, that's just one part of mathematical logic."

"Why bother, though?" Yuri asked. "What's all the fuss about this formalism stuff?"

"Rigor," Miruka said. "You can't have a rigorous discussion without formalism. Say you wanted to claim that a proof was impossible. In that case you'd have to start out by defining what a proof is and what it means for a proof to be impossible. After all, maybe it's not impossible. Maybe you just weren't able to find a proof yourself. Without those definitions you couldn't be sure."

Many times in the past I'd found myself wanting reassurance that it was my own inability to prove something that was getting in my way. That a proof was out there, I just hadn't found it yet.

Miruka continued. "Formalism is also a kind of objectification. It makes what it is you want to talk about exactly clear. The study of mathematics using mathematics itself is called 'metamathematics.' Formal systems put mathematics in a form that allows you to do that."

10.5.2 Math on Math

"So Miruka," Yuri said. "I heard once about how Gödel's incompleteness theorems say something like how life is fun because it's incomplete. Because if you knew everything, nothing would be interesting."

"I guess there are people who think that way," Miruka said with a wry smile. "And sure, there can be beauty in incompleteness . . ." She looked up and closed her eyes. "Like when you consider a beautiful piece of lace and think about how the holes are part of the design. If you don't consider the holes, you can't understand the pattern—you're just considering the surface of things. You aren't seeing the structure, which is where the deeper enjoyment lies."

Miruka opened her eyes and looked at Yuri.

"Formalizing mathematics lets us study its intricate lacework. That's what it means to do math on math—looking at the structure of a theory you're interested in, finding relationships between different theories . . ."

"Resolving Galileo's doubts," I suggested.

Miruka nodded.

"So in that sense, incompleteness isn't a failure or a fault. It's the gateway to new worlds."

10.5.3 Rude Awakening

After lunch I broke away to find a vending machine so I'd have something to drink during the next round of our study group. When I got back to the room nobody was there, just a note from Yuri on the whiteboard:

> We're out taking a look around the library!
> Back soon! ♡ ♡ ♡

A private tour with Miruka, huh? I felt a twinge of jealousy.

I sipped from my water as I reviewed what Miruka had written on the board. No doubt I was missing bits and pieces, but I was pretty sure I'd gotten the gist of things. The big picture was that we'd created a formal system. Next we needed to define Gödel numbers and something called primitive recursion. I wondered if we would end up doing a proof by contradiction—assuming a formal proof of some statement, then showing that doing so led to a contradiction.

My belly was full and the room was warm and quiet, so of course my eyelids started to droop. As soon as I put my head on the desk I was out like a light. Later, the sound of the door opening and three chattering girls entering the room was enough to only half wake me up.

"Oh, this? It's a fish," said Tetra.

"It looks like a secret code," said Yuri. "Ha! He fell asleep."

"From studying all night, I'll bet."

"Hmph. So what's all this about changing attitudes?" Miruka asked.

"Oh, that?" Tetra said. "Well, I used to think that I was slow, but pretty good about sticking to things until I got them. But I'm starting to think that's not enough for some parts of math. That I also need... what? Inspiration, I guess."

"I *totally* know what you mean," said Yuri.

"It's hard to call up inspiration on demand," continued Tetra, "but I figure maybe I can sort of broaden the way I look at things. Like you guys do. I'm learning so much from you. How to solve problems and stuff, sure, but not just that. I'm learning the right attitude to have when approaching math. That it isn't about getting answers and scoring points on tests—it's about maintaining a balance between seriousness and fun."

"I'm not sure this guy is striking any kind of balance," Yuri said. "All he does is study math!"

"What's he like at home?"

"Kinda... slow to pick up on things."

Hey now!

"He should be nicer to his mom too."

"He should wake up is what he should do," Miruka said.

I felt something cold and wet trickling down my neck. I shouted and jumped up in surprise, and saw Miruka smirking, my water bottle in hand.

"Rise and shine," she said. "It's Summer time."

10.6   SUMMER: GÖDEL NUMBERS

10.6.1   *Gödel Numbers for Elementary Symbols*

"Summer is all about Gödel numbers," Miruka said. "They're numbers we assign to the symbols, sequences of symbols, and sequences of sequences of symbols."

She walked back to the board and grabbed her marker.

"First off, Gödel numbers for our elementary symbols. For those we'll use odd numbers up to 13."

Constant	0	s	¬	∨	∀	(	)
Gödel number	1	3	5	7	9	11	13

"Why odd numbers?" asked Tetra.

"You'll see." Miruka grinned. "We'll use prime numbers greater than 13 for type-1 variables."

Type 1 variable	x_1	y_1	z_1	$\cdots$
Gödel number	17	19	23	$\cdots$

"For type-2 variables, we'll use squares of primes greater than 13."

Type 2 variable	x_2	y_2	z_2	$\cdots$
Gödel number	17^2	19^2	23^2	$\cdots$

"For type-3 variables, we'll use cubes of primes greater than 13."

Type 3 variable	x_3	y_3	z_3	$\cdots$
Gödel number	17^3	19^3	23^3	$\cdots$

"And so on and so on—the Gödel number for a type-n variable will be a prime larger than 13 raised to the n-th power. Good so far?"

We nodded.

"Okay, that's it for the basics, constants and variables. Now we have to deal with them in bunches."

10.6.2 Gödel Numbers for Sequences

"Time to define Gödel numbers for sequences," Miruka said. "These will be finite sequences, remember. Since we've defined Gödel numbers for the elementary symbols, we can represent a sequence of symbols as a sequence of Gödel numbers. Like this."

$$n_1, \; n_2, \; n_3, \; \ldots, \; n_k$$

"And it's easy enough to change that sequence into a product, right?"

$$2^{n_1} \times 3^{n_2} \times 5^{n_3} \times \cdots \times p_k^{n_k}$$

"This product is what we use as the Gödel number for the sequence."

"The p's are prime numbers, right?" Tetra asked.

"Right, so p_k will be the k-th prime number, starting from 2."

"Can we do a quick example?" I asked.

"Sure. Let's find the Gödel number for the numeral 2, which is written in symbols as ss0. The Gödel number for s is 3, and the number for 0 is 1, so we can represent the sequence of elementary symbols ss0 like this."

$$3, \; 3, \; 1$$

"Turn that into a product like we discussed, and we get this."

$$2^3 \times 3^3 \times 5^1$$

"Multiply those together and you get . . . 1080. So that's the Gödel number for the numeral 2."

"Hang on, the Gödel number for the numeral 2 is 1080?" Yuri asked. "Why not just 2?"

"We said that in the formal world we're going to represent the number 2 from the world of meaning as the numeral ss0, right?"

"Right."

"So we're looking for the Gödel number for the *string of symbols* ss0, not for a *number.*"

"Ah, okay. I guess that makes sense, then."

"Tetra, now do you see why we're using odd numbers?"

"Uh . . . no?"

"So that we can use parity to check for sequences."

"Oh, I get it! If a Gödel number is even, it's the number for a sequence!"

Miruka nodded.

"The example we just did was finding the Gödel number for a sequence of symbols, but you can create the number for a sequence of sequences of symbols in the same way—use the Gödel numbers you find for each sequence of symbols as exponents of primes, then take the product. The unique factorization theorem guarantees that you can always reconstitute the sequence from its number.

"Back and forth as you please," I said. "Very cool."

"Indeed. This method of prime exponentiation is what Gödel used in his paper, but the method itself isn't important. Do anything you like, so long as that back-and-forth is possible."

Miruka paused to take a swig of my water, then to pop another chocolate into her mouth.

"Now we're ready to generate Gödel numbers for propositional formulas," she said. "They're just sequences of symbols, after all. And formal proofs are just sequences of formulas, so we can create a Gödel number for those, too. In other words, everything in our formal system can be represented by Gödel numbers."

"Can you distinguish between them, though?" Tetra asked. "Between formulas and proofs, I mean."

"I was just about to ask you the same thing," Miruka replied.

"Um, I'm not sure. They'll both be even numbers, so . . . "

"Oh, I see," I began, but Miruka shushed me.

Tetra's face brightened.

"I get it! It's the number of 2's after you do prime factorization!"

"What about them?"

"If you get an odd number of 2's it's a sequence of symbols, but if there's an even number of 2's it's a sequence of sequences!"

"Well done."

Tetra blushed.

"The incompleteness theorems are in essence about whether certain formal proofs exist within formal systems," Miruka said. "Of course, that's a nonsense question if the formal system you're looking at isn't one that can handle formal proofs. That's why we're doing all this, setting things up so that we can numerically encode proofs as Gödel numbers—if the system is powerful enough to handle numbers, then it's powerful enough to deal with formal proofs."

"So it acts a lot like a computer, using bits to represent information?" Tetra asked.

"Yes, but actually it's more that computers act like formal systems," Miruka said. "Don't forget, the first digital computers appeared in the 1940s. That was a post-Gödel world they were born into."

10.7 Fall: Primitive Recursion

10.7.1 Primitive Recursive Functions

"Summer has come to a close, and now Fall is upon us," Miruka said. "We're going to take a break from the world of our formal system P and spend some time in the world of meaning. That's where we'll define something called a primitive recursive function."

"Is that like a function from normal math?" Yuri asked.

"Something like that. Specifically, they're functions that repeat themselves up to some number of repetitions required to find their value."

"Uh..."

"Factorials are a good example," Miruka said, moving back to the whiteboard. "We define the factorial of n as $n! = n \times (n-1) \times \cdots \times 1$, which as a primitive recursive function we call $\texttt{factorial}(n)$, defined like this."

$$\begin{cases} \texttt{factorial}(0) & = 1 \\ \texttt{factorial}(n+1) & = (n+1) \times \texttt{factorial}(n) \end{cases}$$

"Let's try finding factorial(3)"

factorial(3)

$= \underbrace{(2+1)} \times$ factorial(2)	by definition, with $n = 2$
$= (2+1) \times \underbrace{(1+1)} \times$ factorial(1)	by definition, with $n = 1$
$= (2+1) \times (1+1) \times \underbrace{(0+1)} \times$ factorial(0)	by definition, with $n = 0$
$= (2+1) \times (1+1) \times (0+1) \times 1$	by definition of factorial(0)
$= 3 \times 2 \times 1 \times 1$	calculate
$= 6$	calculate

"So to calculate factorial(3), we have to use the definition 4 times. In general, to calculate factorial(n) we'll have to use the definition $n + 1$ times. That's what it means to say there's some upper limit on the number of repetitions."

"We're using addition and multiplication to get the final value, though." I said. "Did we ever define those?"

"No we didn't, but we will. So let's take a step back and do things right, to get ready for that. Let F, G, and H be functions that take natural number arguments, including 0. Then if we define F like this, we say it's a function that is defined by primitive recursion of functions G and H."

$$\begin{cases} F(0, \quad x) &=& G(x) \\ F(n+1, \quad x) &=& H(n, x, F(n, x)) \end{cases}$$

"For example, in the case of factorial(n), we say that $F(n, x) =$ factorial(n), $G(x) = 1$, and $H(n, x, y) = (n + 1) \times y$. Then $F(n, x)$ is defined by primitive recursion of $G(x)$ and $H(n, x, y)$."

"Wow, this is getting kind of hard to follow," Tetra said.

"Maybe writing out $F(3, x)$ will make things clearer."

$$F(3, x) = \underline{H(2, x, F(2, x))}$$
$$= H(2, x, \underline{H(1, x, F(1, x))})$$
$$= H(2, x, H(1, x, \underline{H(0, x, F(0, x))}))$$
$$= H(2, x, H(1, x, H(0, x, \underline{G(x)})))$$

"See? We can find $F(n, x)$ by using function G once and function H n times. What we've talked about here is only good for two variables, but we can extend it to N variables by defining it like this."

$$\begin{cases} F(0, & \vec{x}) & = & G(\vec{x}) \\ F(n+1, & \vec{x}) & = & H(n, \vec{x}, F(n, \vec{x})) \end{cases}$$

I raised my eyebrows. "Vectors?"

"Yes, but probably not like you're thinking. I'm using $\vec{x}$ as an abbreviation for the sequence $x_1, x_2, \ldots, x_{N-1}$."

"Ah, okay."

"So anyway that's the basic method, and here's what we want to do with it—define what kinds of functions are primitive recursive."

▷ **PRF-1**

Constant functions are primitive recursive.

▷ **PRF-2**

The function that returns the successor of its argument is primitive recursive.

▷ **PRF-3**

A function that is defined by primitive recursion of two primitive recursive functions is primitive recursive.

▷ **PRF-4**

A function that returns its argument with variables replaced by primitive recursive functions is primitive recursive.

▷ **PRF-5**

Projection functions like $F(\vec{x}) = x_k$ that return a single extracted variable are primitive recursive.

"Now we can use primitive recursive functions to define primitive recursive predicates."

▷ **Primitive recursive predicates**

A predicate $R(n, \vec{x})$ is called a *primitive recursive predicate* if there exists a primitive recursive function $F(n, \vec{x})$ such that

$$R(n, \vec{x}) \iff F(n, \vec{x}) = 0.$$

"Miruka, I'm totally lost," Tetra said.

"Lost how?"

"Like, what are we doing here?"

"Defining primitive recursive predicates."

"That's ... not super helpful."

"Don't overthink it. We're just defining a kind of predicate that has some restrictions on it."

"And why are we doing that?"

"Because we'll need it to prove the incompleteness theorems, of course."

10.7.2 Properties of Primitive Recursive Functions and Predicates

"Speaking of which," Miruka said, "we're also going to need a few theorems related to primitive recursive functions and predicates."

▷ **PRT-1**

Any function (predicate) obtained by substituting primitive recursive functions in place of variables in another primitive recursive function (predicate) is itself primitive recursive.

▷ **PRT-2**

If R and S are both primitive recursive relations, then $\neg R$ and $R \vee S$ (and therefore also $R \wedge S$) are primitive recursive.

▷ **PRT-3**

If F and G are both primitive recursive functions, then the predicate F = G is primitive recursive.

▷ **PRT-4**

If M is a primitive recursive function and R is a primitive recursive predicate, then the following S is a primitive recursive predicate.

$$S(\vec{x}, \vec{y}) \iff \forall n \left[n \leqslant M(\vec{x}) \Rightarrow R(n, \vec{y}) \right]$$

In other words, $S(\vec{x}, \vec{y})$ holds if and only if $R(n, \vec{y})$ holds for all values of n less than or equal to $M(\vec{x})$. Here, $M(\vec{x})$ is an upper limit, and $\vec{x}$ and $\vec{y}$ are each finite sequences of variables.

▷ **PRT-5**

If M is a primitive recursive function and R is a primitive recursive predicate, then the following T is a primitive recursive predicate.

$$T(\vec{x}, \vec{y}) \iff \exists n \left[n \leqslant M(\vec{x}) \land R(n, \vec{y}) \right]$$

In other words, there exists some value of n less than or equal to $M(\vec{x})$ such that $R(n, \vec{y})$ holds. Here, $M(\vec{x})$ is an upper limit.

▷ **PRT-6**

If M is a primitive recursive function and R is a primitive recursive predicate, then the following F is a primitive recursive function.

$$F(\vec{x}, \vec{y}) = \min n \left[n \leqslant M(\vec{x}) \land R(n, \vec{y}) \right]$$

This function returns the smallest value of n less than or equal to $M(x)$ for which $R(n, y)$ holds. If there exists no such value of n, this function returns 0. Here, $M(\vec{x})$ is an upper limit.

Tetra let out a small moan.

"What's wrong?" Miruka asked.

"Too many words..." Tetra said. "I think my head is about to explode."

"I doubt that."

"Just give me a little time to let all this sink in, please."

"Okay, but I hope you see that most of this isn't new stuff. We're just being a lot more specific about things we use all the time."

"Even me?" Yuri asked.

"Sure. Addition and subtraction, for example. Exponents, equality and inequality . . . Those are all primitive recursive. When Winter comes we'll actually build our own primitive recursive functions and predicates, one by one. It'll be fun."

"But I can't see at all what this has to do with the incompleteness theorems," Tetra said.

"Hmph. Well then, let's talk about a very important theorem that all primitive recursive predicates have to uphold."

10.7.3 The Representation Theorems

"The proof of the incompleteness theorems uses something I'm going to call the representation theorem," Miruka said. "Since our formal system P describes a number theory, this representation theorem has to hold. I'm going to present it using just two variables, but trust me when I say that it will hold just as well for any number of variables."

Representation theorem

If R is a bivariate primitive recursive predicate, then for all numbers m, n there exists a bivariate formula r such that the following both hold.

▶ FALL-1

$R(m, n) \Rightarrow$ There exists a $\langle\!\langle$formal proof$\rangle\!\rangle$ of $r\langle\overline{m}, \overline{n}\rangle$

▶ FALL-2

$\neg R(m, n) \Rightarrow$ There exists a $\langle\!\langle$formal proof$\rangle\!\rangle$ of $\mathrm{not}(r\langle\overline{m}, \overline{n}\rangle)$

In this case, we say that the formula r *numerically identifies* the predicate R.

"The representation theorem guarantees the existence of some r that represents R. The predicate R is a concept from the world of meaning. The formula r is a concept from the world of formalism. In other words, this representation theorem is our bridge between the two worlds, and primitive recursion is the passport that lets us cross over that bridge."

"I don't get what this $r\langle \overline{m}, \overline{n} \rangle$ means," Yuri said.

"Then let's talk about that," Miruka said.

▷ **Predicates and propositions**

A predicate R having two free variables, into which numbers m, n are substituted, is represented as the proposition $R(m, n)$.

"So this is how we differentiate between predicates and propositions?" Tetra asked.

"Right," said Miruka. "A predicate is a statement like 'y evenly divides x.' It has free variables, so as it is we can't say anything about whether it holds or not."

"Got it. And we insert specific values into the free variables to turn it into a proposition, right?"

"Exactly. Then we can talk about how the proposition '3 evenly divides 12' holds, while the proposition '7 evenly divides 12' doesn't."

▷ **Propositional formulas and statements**

A ⟨⟨propositional formula⟩⟩ r having two ⟨⟨free variables⟩⟩, into which ⟨⟨numerals⟩⟩ $\overline{m}, \overline{n}$ are substituted, is represented as the ⟨⟨statement⟩⟩ $r\langle \overline{m}, \overline{n} \rangle$.

"So this time we're differentiating between propositional formulas and statements," Tetra said.

"Yep. Remember, a 'statement' is a formula that has no free variables."

"And predicates and propositions are concepts from the world of meaning, while propositional formulas and statements are concepts from the world of formalism?"

"I was just about to ask you that, but you beat me to it."

"So what's r?" I asked. "Is that a formula, or the Gödel number for a formula?"

"The latter. Specifically, $r\langle\overline{m}, \overline{n}\rangle$ is the Gödel number for a statement."

"Help me make sure I've got all this straight," Tetra said, scanning the notes she had made. "So the representation theorem connects the world of meaning and the world of formalism. And the existence of a small-r that represents a big-R means, let's see... If the proposition $R(m, n)$ holds, then somewhere out there is a formal proof of the statement $r\langle\overline{m}, \overline{n}\rangle$. But if the proposition $R(m, n)$ *doesn't* hold, then there's no way to prove that statement?"

"The second part there isn't quite right," Miruka said. "Read the theorem closely."

"Huh? I thought... Oh! Proposition $R(m, n)$ not being true doesn't mean you can't prove $r\langle\overline{m}, \overline{n}\rangle$, it means you *can* prove its negation."

"There ya go."

Tetra cocked her head in thought.

"Maybe I'm missing something," she said, "but isn't it kind of obvious that formulas represent predicates?"

"Not necessarily," Miruka said. "Sure, you can always represent a predicate in the world of meaning. But the representation theorem is asserting something much stronger for primitive recursive predicates in particular. It says that when you create a proposition by substituting the variables in a predicate with actual numbers, whether or not that proposition holds *can be determined using a formal proof*. Think about that—it means we can talk about the truth of things in the world of meaning, from the viewpoint of the world of formalism. So 'representation' in this sense is pretty powerful stuff. But keep in mind the requirement for the predicate to be primitive recursive. That means predicates that don't have an upper limit—$\forall$ and $\exists$—won't necessarily have formulas that represent them in the world of meaning."

"I still don't get all this primitive recursive stuff," Yuri said.

"It'll come, just give it time. In the meantime, let's move on to the next season."

"If you say so..."

"From Fall to Winter," Miruka said in a singsong voice. "Winter will keep us busy defining predicates to use in our formal system. And if we can give those definitions as primitive recursive predicates..."

Miruka raised an eyebrow at Tetra.

"Um... then we know there's a formula that represents that predicate?" Tetra said.

"Exactly. Put more precisely, if a predicate in our formal system is primitive recursive, then we can be sure there exists *within that system itself* a propositional formula that represents the predicate. And why can we be sure of that?"

"From the representation theorem."

Miruka nodded, then spoke in hushed tones.

"It's going to be a slog, but we'll be trudging through the snow with a single goal in mind—we're after a specific, very powerful predicate."

10.8 WINTER: THE LONG, HARD JOURNEY TO PROVABILITY

10.8.1 *Gearing Up*

"Our goal in Winter is to create a primitive recursive predicate that says 'p is a formal proof of x.' But before we head out, we have to gear up a bit."

Miruka went back to the whiteboard and started writing as she talked.

"First off, we're going to use 'so-and-so holds for any x where $x \leqslant M$' a lot, so we'll write that like this."

$$\forall x \leqslant M \left[\text{something} \right] \overset{\text{def}}{\Longleftrightarrow} \forall x \left[x \leqslant M \Rightarrow \text{something} \right]$$

"The 'def' stands for 'define'?" I asked.

"Defining a predicate, yeah," Miruka said. "We also need something similar for 'there exists an x less than or equal to M,' so let's write that like this."

$$\exists x \leqslant M \left[\text{something} \right] \overset{\text{def}}{\Longleftrightarrow} \exists x \left[x \leqslant M \wedge \text{something} \right]$$

"And one more, to define a function that returns the minimum x that makes such-and-such a condition true."

$$\min x \leqslant M \left[\text{something} \right] \overset{\text{def}}{=} \min x \left[x \leqslant M \wedge \text{something} \right]$$

"We'll say that if no such x exists, the function will return 0."

"And here the $\overset{\text{def}}{=}$ symbol means we're defining a function, right?" Tetra said.

"Right. Oh, one more thing we need to define, to make things easier to read—Gödel numbers for our seven elementary symbols."

$$\boxed{0} \overset{\text{def}}{=} 1 \quad \boxed{s} \overset{\text{def}}{=} 3 \quad \boxed{\neg} \overset{\text{def}}{=} 5 \quad \boxed{\vee} \overset{\text{def}}{=} 7$$

$$\boxed{\forall} \overset{\text{def}}{=} 9 \quad \boxed{(} \overset{\text{def}}{=} 11 \quad \boxed{)} \overset{\text{def}}{=} 13$$

"Okay, I think we're ready to set out on our journey. We'll be making 46 stops on our tour of the world of meaning, each one a definition that leads us closer to our destination."

10.8.2 Definitions from Number Theory

"Here's our first definition in the world of meaning," Miruka said.

> **Definition 1.** CanDivide(x, d) *is a predicate meaning 'd evenly divides x.'*
>
> $$\text{CanDivide}(x, d) \overset{\text{def}}{\iff} \exists n \leqslant x \left[x = d \times n \right]$$

"This says that there exists some n less than or equal to x such that $x = d \times n$."

"Ah, interesting," I said. "So we can say '3 evenly divides 12' by writing CanDivide$(12, 3)$."

$$\exists n \leqslant 12 \left[12 = 3 \times n \right]$$

"And in this case n would be 4?" Tetra said.

"Right. And that means CanDivide$(12, 3)$ holds."

"So this is an expression of the possibility of existence, too."

"What do you mean?" Yuri asked.

"That we're representing the idea of being divisible as a statement about existence."

"You're the only person I know who would notice things like that," I said.

Definition 2. IsPrime(x) *is a predicate meaning 'x is a prime number.'*

$$\text{IsPrime}(x) \overset{\text{def}}{\Longleftrightarrow} x > 1 \wedge$$
$$\neg\Big(\exists\, d \leqslant x \Big[\, d \neq 1 \wedge d \neq x \wedge \text{CanDivide}(x, d)\,\Big]\Big)$$

"Take a shot at reading that, Yuri," Miruka said.

"Okay, uh... remind me what CanDivide(x, d) is again?"

"It says that d evenly divides x," I said.

"Oh, so this says there's no d less than or equal to x that can evenly divide x?"

"You're forgetting the part about d not equalling 1 or x."

"I didn't forget it, I just hadn't gotten to that part yet."

"There's also the condition that x > 1. So yeah, this looks like the definition of a prime number."

Definition 3. prime(n, x) *is a fuction that returns the n-th prime factor of x. We assume that the primes are ordered from smallest to largest. For convenience, we define the 0-th prime to be 0.*

$$\begin{cases} \text{prime}(0, x) \overset{\text{def}}{=} 0 \\ \text{prime}(n + 1, x) \overset{\text{def}}{=} \min p \leqslant x \Big[\, \text{prime}(n, x) < p \wedge \\ \qquad\qquad\qquad\qquad\qquad \text{CanDivideByPrime}(x, p)\,\Big] \end{cases}$$

where CanDivideByPrime(x, p) *is defined as*

$$\text{CanDivideByPrime}(x, p) \overset{\text{def}}{\Longleftrightarrow} \text{CanDivide}(x, p) \wedge \text{IsPrime}(p).$$

"So this returns, let's see... the smallest prime that evenly divides x and is larger than the n-th prime factor of x, I guess."

"Example, please," Yuri said.

"For example $2^4 \times 3^1 \times 7^2 = 2352$, so you get this."

$\text{prime}(0, 2352) = 0$ from the definition

$\text{prime}(1, 2352) = 2$ smallest prime $> \text{prime}(0, 2352)$ that divides 2352

$\text{prime}(2, 2352) = 3$ smallest prime $> \text{prime}(1, 2352)$ that divides 2352

$\text{prime}(3, 2352) = 7$ smallest prime $> \text{prime}(2, 2352)$ that divides 2352

"Much clearer, thanks," Yuri said.

Definition 4. $\text{factorial}(n)$ *is a fuction that returns the factorial of* n.

$$\begin{cases} \text{factorial}(0) & \overset{\text{def}}{=} 1 \\ \text{factorial}(n+1) & \overset{\text{def}}{=} (n+1) \times \text{factorial}(n) \end{cases}$$

Definition 5. p_n *is a fuction that returns the* n-*th prime. For convenience, we define the* 0-*th prime to be* 0.

$$\begin{cases} p_0 & \overset{\text{def}}{=} 0 \\ p_{n+1} & \overset{\text{def}}{=} \min p \leqslant M_5(n) \Big[p_n < p \wedge \text{IsPrime}(p) \Big] \end{cases}$$

where $M_5(n)$ *is defined as*

$$M_5(n) \overset{\text{def}}{=} \text{factorial}(p_n) + 1.$$

"The n-th prime?" Yuri said.

"Right, so $p_0 = 0, p_1 = 2, p_2 = 3, p_3 = 5, p_4 = 7$, and so on."

"Where does that $p \leqslant M_5(n)$ come from?"

"Because $M_5(n) = \text{factorial}(p_n) + 1 = 1 \times 2 \times 3 \times \cdots \times p_n + 1$."

"So?"

"Well, $M_5(n)$ is greater than p_n, and we know that there has to be a prime equal to or less than $M_5(n)$."

"And?"

"And that's why if we're going to search for p_{n+1} we need to stay at or below $M_5(n)$."

10.8.3 Definitions for Sequences

"Next we introduce sequences. We'll do that using prime exponentiation. Here's the first definition."

> **Definition 6.** $x[n]$ *is a fuction that returns the n-th term of sequence x. Note that we assume $1 \leqslant n \leqslant$ sequence length.*
>
> $$x[n] \stackrel{\text{def}}{=} \min k \leqslant x \Big[\text{CanDivideByPower}(x, n, k) \land$$
> $$\neg \text{CanDivideByPower}(x, n, k+1) \Big]$$
>
> *where* $\text{CanDivideByPower}(x, n, k)$ *is defined as*
>
> $$\text{CanDivideByPower}(x, n, k) \stackrel{\text{def}}{\iff} \text{CanDivide}(x, \text{prime}(n, x)^k).$$

"I'm not sure I get this definition of $\text{CanDivideByPower}(x, n, k)$," Tetra said.

I looked closely at the definition, rearranging things in my head.

"It says that x can be evenly divided by the k-th power of $\text{prime}(n, x) \ldots$ I think?"

"But what does that have to do with $x[n]$?"

"Hmm ... Oh, it's saying that x is evenly divisible by the k-th power of $\text{prime}(n, x)$ but not by the $\{k + 1\}$-th power. Right, right. That means x has exactly k $\text{prime}(n, x)$'s as prime factors."

"Huh?"

"That means the exponent k on $\text{prime}(n, x)$ will be the n-th term in the sequence x."

> **Definition 7.** $\text{len}(x)$ *is a fuction that returns the length of sequence x.*
>
> $$\text{len}(x) \stackrel{\text{def}}{=} \min k \leqslant x \Big[\text{prime}(k, x) > 0 \land \text{prime}(k+1, x) = 0 \Big]$$

"This lets you get the first term in the sequence x as $x[1]$, and the last term as $x[\text{len}(x)]$," Miruka said. "For example if you let x

be the sequence $\boxed{\forall}\,\boxed{x_1}\,\boxed{(}\;\cdots\;\boxed{)}$, you get this."

$$
\begin{array}{ccccc}
x[1] & x[2] & x[3] & \cdots & x[\mathrm{len}(x)] \\[4pt]
\boxed{\forall} & \boxed{x_1} & \boxed{(} & \cdots & \boxed{)} \\[6pt]
\| & \| & \| & \cdots & \| \\[4pt]
9 & 17 & 11 & \cdots & 13
\end{array}
$$

Definition 8. $x * y$ *is a fuction that returns the concatenation of sequences* x *and* y.

$$
x * y \overset{\text{def}}{=} \min z \leqslant M_8(x,y)
$$
$$
\Big[\, \forall m \leqslant \mathrm{len}(x)\,\Big[\,1 \leqslant m \Rightarrow z[m] = x[m]\,\Big]
$$
$$
\wedge \forall n \leqslant \mathrm{len}(y)\,\Big[\,1 \leqslant n \Rightarrow z[\mathrm{len}(x)+n] = y[n]\,\Big]\,\Big]
$$

where $M_8(x,y)$ *is defined as*

$$
M_8(x,y) \overset{\text{def}}{=} (p_{\mathrm{len}(x)+\mathrm{len}(y)})^{x+y}.
$$

"This one says something like this. Say you have sequence z with terms $z[1]$ through $z[\mathrm{len}(x)]$ equal to their corresponding terms in sequence x, and terms $z[\mathrm{len}(x)+1]$ through $z[\mathrm{len}(x)+\mathrm{len}(y)]$ equal to the terms in sequence y. Then z is the concatenation of x and y."

$$
\begin{array}{cccccc}
x[1] & \cdots & x[\mathrm{len}(x)] & y[1] & \cdots & y[\mathrm{len}(y)] \\[6pt]
\| & \cdots & \| & \| & \cdots & \| \\[6pt]
z[1] & \cdots & z[\mathrm{len}(x)] & z[\mathrm{len}(x)+1] & \cdots & z[\mathrm{len}(x)+\mathrm{len}(y)]
\end{array}
$$

Definition 9. $\langle x \rangle$ *is a fuction that returns a sequence created from just* x. *Here,* $x > 0$.

$$
\langle x \rangle \overset{\text{def}}{=} 2^x
$$

Definition 10. paren(x) *is a fuction that returns a sequence with* x *in parentheses.*

$$\mathsf{paren}(x) \overset{\text{def}}{=} \langle\; \boxed{(} \;\rangle * x * \langle\; \boxed{)} \;\rangle$$

"You can read this one, right Yuri?" Miruka said.

"Umm... You're just connecting together $\boxed{(}$ and x and $\boxed{)}$."

"Right. You can call this 'parenthesizing.'"

10.8.4 Variables, Symbols, and Formulas

"Time to add variables into the mix," Miruka said.

Definition 11. IsVarType(x, n) *is a predicate meaning '*x *is a* $\langle\!\langle type\text{-}n \rangle\!\rangle$ $\langle\!\langle variable \rangle\!\rangle$.'

$$\mathsf{IsVarType}(x, n) \overset{\text{def}}{\Longleftrightarrow} n \geqslant 1 \wedge \exists p \leqslant x \left[\, \mathsf{IsVarBase}(p) \wedge x = p^n \,\right]$$

where IsVarBase(p) *is defined as*

$$\mathsf{IsVarBase}(p) \overset{\text{def}}{\Longleftrightarrow} p > \boxed{)} \wedge \mathsf{IsPrime}(p).$$

"I've been wondering," I said. "What's with the brackets you're putting around some words?"

"I'm using those to indicate words that are concepts from meta-mathematics."

"Oh?"

"Remember, the x here doesn't indicate a variable in the world of meaning. It's a Gödel number for a variable that was defined in our formal system P."

"What does $p > \boxed{)}$ mean?" Tetra asked.

"It means $p > 13$. Don't forget how we assigned Gödel numbers to our variables."[3]

---

[3] see p. 301

Definition 12. IsVar(x) *is a predicate meaning 'x is a* $\langle\!\langle variable \rangle\!\rangle$.'

$$\text{IsVar}(x) \stackrel{\text{def}}{\Longleftrightarrow} \exists n \leqslant x \left[\text{IsVarType}(x, n) \right]$$

"Can you read this one, Yuri?" Miruka asked.

"It says there's an n that makes x into a type-n variable."

"Ah, and if there exists an n that makes x a type-n variable, then x is a variable," I said.

"Let's add some logic operators," Miruka said.

Definition 13. not(x) *is a fuction that returns* $\langle\!\langle \neg(x) \rangle\!\rangle$.

$$\text{not}(x) \stackrel{\text{def}}{=} \langle\, \boxed{\neg}\, \rangle * \text{paren}(x)$$

"The $\langle\, \boxed{\neg}\, \rangle * \text{paren}(x)$ part would be the logic statement $\neg(\cdots)$, huh." I said.

"Hey, we've seen that $\text{not}(x)$ before," Tetra said. "When we talked about the representation theorem."

Definition 14. or(x, y) *is a fuction that returns* $\langle\!\langle (x) \vee (y) \rangle\!\rangle$.

$$\text{or}(x, y) \stackrel{\text{def}}{=} \text{paren}(x) * \langle\, \boxed{\vee}\, \rangle * \text{paren}(y)$$

Definition 15. forall(x, a) *is a fuction that returns* $\langle\!\langle \forall x(a) \rangle\!\rangle$.

$$\text{forall}(x, a) \stackrel{\text{def}}{=} \langle\, \boxed{\forall}\, \rangle * \langle\, x\, \rangle * \text{paren}(a)$$

"So this function takes a variable x and a formula a, and returns $\forall x(a)$," I said.

"More precisely, given a Gödel number x that represents some variable, and a Gödel number a that represents some formula, the function $\text{forall}(x, a)$ returns a Gödel number that represents $\forall x(a)$."

"Ah, of course. In the world of meaning, everything in the formal system P is represented as a number."

"What does that mean, exactly?" Tetra asked.

"That even things like variables and formulas are all Gödel numbers."

"So don't you need to check $\mathrm{IsVar}(x)$?" Yuri asked.

Miruka beamed at her. "I'm glad you noticed," she said. "Yes, you do when you use $\mathrm{forall}(x, a)$."

Definition 16. $\mathrm{succ}(n, x)$ *is a fuction that returns the* n-*th successor of* x.

$$\begin{cases} \mathrm{succ}(0, x) & \overset{\text{def}}{=} x \\ \mathrm{succ}(n+1, x) & \overset{\text{def}}{=} \langle\, \boxed{\text{s}}\, \rangle * \mathrm{succ}(n, x) \end{cases}$$

"So $\mathrm{succ}(0, x)$ would just be x?" Yuri asked.

"Seems that way," I said. "I'm not sure how else you would define the 0th successor of a variable."

"And in the $\mathrm{succ}(n+1, x)$ case, we're getting s concatenated with $\mathrm{succ}(n, x)$?" Tetra asked.

"Right. The function calls itself over and over again."

"But with $n+1$ decreasing by 1."

"Until it reaches $\mathrm{succ}(0, x)$, which stops it from going any lower."

"Didn't we already see something like this?" Yuri said.

"Let's see... Oh, right. In definition 4, for the $\mathrm{factorial}(n)$ function."

Definition 17. $\overline{n}$ *is a fuction that returns the* $\langle\!\langle numeral \rangle\!\rangle$ *for* n.

$$\overline{n} \overset{\text{def}}{=} \mathrm{succ}(n, \langle\, \boxed{0}\, \rangle)$$

"So this one is defined as the n-th successor of 0," Tetra said.

"Which means $\overline{n}$ gives the Gödel number for n s's, plus a zero."

$$\underbrace{\text{ss}\cdots\text{s}}_{n \text{ of these}} 0$$

Definition 18. IsNumberType(x) *is a predicate meaning* 'x *is a* ⟨⟨*type*-1 *variable*⟩⟩.'

$$\text{IsNumberType}(x) \stackrel{\text{def}}{\Longleftrightarrow}$$

$$\exists\, m, n \leqslant x \left[\left(m = \boxed{0}\ \vee\ \text{IsVarType}(m, 1)\right) \wedge x = \text{succ}(n, \langle m \rangle) \right]$$

"Who wants to read the $m = \boxed{0}\ \vee\ \text{IsVarType}(m, 1)$ part?" Miruka asked.

"Well, I think this $m = \boxed{0}$ here corresponds to something like sss0," I said.

"And the IsVarType(m, 1) part would be sssx₁," Tetra said.[4]

Definition 19. IsNthType(x, n) *is a predicate meaning* 'x *is a* ⟨⟨*type*-n *symbol*⟩⟩.'

$$\text{IsNthType}(x, n) \stackrel{\text{def}}{\Longleftrightarrow} \left(n = 1 \wedge \text{IsNumberType}(x) \right)$$

$$\vee \left(n > 1 \wedge \exists\, v \leqslant x \left[\text{IsVarType}(v, n) \wedge x = \langle v \rangle \right] \right)$$

"Now things are really starting to look like computer programming," Tetra said.

"How's that?" I asked.

"Because we're separating out the cases where $n = 1$ and where $n > 1$."

"We're moving on to something new with the next definition," Miruka said. "Propositional formulas."

---

[4]see p. 292

Definition 20. IsElementForm(x) *is a predicate meaning 'x is an $\langle\!\langle$ elementary propositional formula $\rangle\!\rangle$.'*

$$\text{IsElementForm}(x) \overset{\text{def}}{\iff} \exists\, a, b, n \leqslant x \left[\text{IsNthType}(a, n+1) \right.$$
$$\wedge\, \text{IsNthType}(b, n)$$
$$\left. \wedge\, x = a * \text{paren}(b) \right]$$

where $\exists\, a, b, n \leqslant x \left[\cdots \right]$ *is defined as*

$$\exists\, a, b, n \leqslant x \left[\cdots \right] \overset{\text{def}}{\iff} \exists\, a \leqslant x \left[\exists\, b \leqslant x \left[\exists\, n \leqslant x \left[\cdots \right] \right] \right].$$

"We said that a propositional formula would be something in the form $a(b)$, right?" I said.

"Right," Miruka said, "so long as a is of type-$\{n+1\}$, and b is of type-n."

"Ah, okay. So that's why we check IsNthType$(a, n+1)$ and IsNthType(b, n)."

Definition 21. IsOp(x, a, b) *is a predicate meaning 'x is $\langle\!\langle \neg(a) \rangle\!\rangle$ or $\langle\!\langle (a) \vee (b) \rangle\!\rangle$ or $\langle\!\langle \forall v(a) \rangle\!\rangle$.'*

$$\text{IsOp}(x, a, b) \overset{\text{def}}{\iff} \text{IsNotOp}(x, a) \vee$$
$$\text{IsOrOp}(x, a, b) \vee$$
$$\text{IsForallOp}(x, a)$$

where IsNotOp(x, a), IsOrOp(x, a, b), IsForallOp(x, a) *are respectively defined as*

$$\text{IsNotOp}(x, a) \overset{\text{def}}{\iff} x = \text{not}(a),$$
$$\text{IsOrOp}(x, a, b) \overset{\text{def}}{\iff} x = \text{or}(a, b),$$
$$\text{IsForallOp}(x, a) \overset{\text{def}}{\iff} \exists\, v \leqslant x \left[\text{IsVar}(v) \wedge x = \text{forall}(v, a) \right].$$

"What does Op stand for?" Yuri asked.

"Operator," Miruka said.

"Which means?"

"In this case, one of $\neg$, $\vee$, or $\forall$."

Definition 22. IsFormSeq(x) *is a predicate meaning 'x is a sequence of $\langle\!\langle$propositional formulas$\rangle\!\rangle$ that are each either an $\langle\!\langle$elementary propositional formula$\rangle\!\rangle$ or obtained from preceding formulas through their negation, disjunction, or generalization[a]..'*

$$\text{IsFormSeq}(x) \overset{\text{def}}{\Longleftrightarrow} \text{len}(x) > 0 \wedge$$

$$\forall n \leqslant \text{len}(x) \left[n > 0 \Rightarrow \text{IsElementForm}(x[n]) \vee \right.$$

$$\left. \exists p, q < n \left[p, q > 0 \wedge \text{IsOp}(x[n], x[p], x[q]) \right] \right]$$

---

[a]see F-2 through F-4 on p. 293

"This is a long one, but it's not so hard if you read it carefully," Miruka said.

"$x[n]$ means the n-th term in the sequence x, yeah?" Yuri said.

"And everything in the sequence x is an elementary formula, or ..." Tetra said.

"What's this IsOp($x[n], x[p], x[q]$)?" I asked.

"Be sure to pay attention to the $p, q < n$," Miruka said.

"Oh, of course. $x[n]$ is created from $x[p]$ and $x[q]$."

"Created how?" Tetra asked.

"$x[n]$ is the n-th formula in the sequence, but that was created previously by $x[p]$ and $x[q]$," Miruka said.

"As a formal proof?"

"No, as a definition for a formula. A formula is just a sequence of elementary formulas in the form $\neg(a)$ or $(a) \vee (b)$ or $\forall x(a)$, right?"

"Right."

"So here we're using a sequence of formulas to show that we've followed the process of piecing together elementary formulas, just as the definition of a formula[5] says we should."

"All this thinking is making me hungry," Yuri said.

"Feel free to snack away," Miruka said. "These chocolates are mine, though."

---

[5]see p. 293

Miruka pulled the box of chocolates closer to her and popped one into her mouth.

"Time to introduce bound variables," she said.

Definition 23. $\mathrm{IsForm}(x)$ *is a predicate meaning 'x is a* ⟨⟨*propositional formula*⟩⟩*.' It defines the existence of a* ⟨⟨*sequence of formulas*⟩⟩ n *created from* ⟨⟨*elementary formulas*⟩⟩ *such that x is the final term.*

$$\mathrm{IsForm}(x) \overset{\text{def}}{\Longleftrightarrow} \exists n \leqslant M_{23}(x) \left[\mathrm{IsFormSeq}(n) \wedge \mathrm{IsEndedWith}(n, x) \right]$$

where $M_{23}(x)$ *and* $\mathrm{IsEndedWith}(n, x)$ *are respectively defined as*

$$M_{23}(x) \overset{\text{def}}{=} \left(p_{\mathrm{len}(x)^2} \right)^{x \times \mathrm{len}(x)^2},$$

$$\mathrm{IsEndedWith}(n, x) \overset{\text{def}}{\Longleftrightarrow} n[\mathrm{len}(n)] = x.$$

Definition 24. $\mathrm{IsBoundAt}(v, n, x)$ *is a predicate meaning 'the* ⟨⟨*variable*⟩⟩ v *is* ⟨⟨*bound*⟩⟩ *in x at the n-th position.'*

$$\mathrm{IsBoundAt}(v, n, x) \overset{\text{def}}{\Longleftrightarrow} \mathrm{IsVar}(v) \wedge \mathrm{IsForm}(x)$$

$$\wedge \, \exists a, b, c \leqslant x \left[x = a * \mathrm{forall}(v, b) * c \right.$$

$$\left. \wedge \, \mathrm{IsForm}(b) \wedge \mathrm{len}(a) + 1 \leqslant n \leqslant \mathrm{len}(a) + \mathrm{len}(\mathrm{forall}(v, b)) \right]$$

"Hey, look!" Yuri said. "It's using $\mathrm{forall}(v, b)$, so it checks $\mathrm{IsVar}(v)$, just like I said."

"Is $\mathrm{len}(a) + 1 \leqslant n \leqslant \mathrm{len}(a) + \mathrm{len}(\mathrm{forall}(v, b))$ some kind of range?" Tetra asked.

"Yes," Miruka said, "the range over which the variable v is bound. It's called the scope. But it isn't necessarily the only place where the variable v can appear."

$$\cdots \quad \overbrace{}^{a} \forall v \underbrace{(\overbrace{\cdots}^{b})}_{\text{scope of } v} \overbrace{\cdots}^{c}$$

"Next we add free variables and substitutions," Miruka said.

Definition 25. IsFreeAt(v, n, x) *is a predicate meaning* '⟨⟨*variable*⟩⟩ v *is not* ⟨⟨*bound*⟩⟩ *in* x *at position* n.'

$$\text{IsFreeAt}(v, n, x) \overset{\text{def}}{\iff} \text{IsVar}(v) \wedge \text{IsForm}(x) \wedge$$
$$v = x[n] \wedge n \leqslant \text{len}(x) \wedge \neg\text{IsBoundAt}(v, n, x)$$

Definition 26. IsFree(v, x) *is a predicate meaning* 'v *is* ⟨⟨*free*⟩⟩ *in* x.'

$$\text{IsFree}(v, x) \overset{\text{def}}{\iff} \exists n \leqslant \text{len}(x) \left[\text{IsFreeAt}(v, n, x) \right]$$

Definition 27. substAtWith(x, n, c) *is a fuction that returns* x *with the term at position* n *replaced by* c. *We assume that* $1 \leqslant n \leqslant \text{len}(x)$.

$$\text{substAtWith}(x, n, c) \overset{\text{def}}{=}$$
$$\min z \leqslant M_8(x, c) \left[\exists a, b \leqslant x \left[n = \text{len}(a) + 1 \right. \right.$$
$$\left. \left. \wedge\, x = a * \langle x[n] \rangle * b \wedge z = a * c * b \right] \right]$$

(Note: see Def. 8 on p. 317 for the definition of the M_8 function.)

"There are so many variables flying around I can hardly tell what's what," Tetra said.

"It helps if you focus on the x's and z's," I said.

$$x \;=\; \overbrace{\cdots}^{a} \;\; x[n] \;\; \overbrace{\cdots}^{b}$$
$$z \;=\; \underbrace{\cdots}_{a} \;\; c \;\; \underbrace{\cdots}_{b}$$

"Okay, so we're saying that z equals the sequence x with the term at position n replaced by c. Got it."

Definition 28. $\mathrm{freepos}(k, \nu, x)$ *is a fuction that returns the* $k + 1st$ *position in* x *where* ν *is* $\langle\!\langle free \rangle\!\rangle$, *as counted backward from the end of the sequence. If there is no position where* ν *is free, this function returns 0.*

$$\mathrm{freepos}(0, \nu, x) \overset{\mathrm{def}}{=} \min n \leqslant \mathrm{len}(x) \Big[\mathrm{IsFreeAt}(\nu, n, x)$$

$$\wedge \neg \Big(\exists p \leqslant \mathrm{len}(x) \Big[n < p \wedge \mathrm{IsFreeAt}(\nu, p, x) \Big] \Big) \Big]$$

$$\mathrm{freepos}(k + 1, \nu, x) \overset{\mathrm{def}}{=} \min n < \mathrm{freepos}(k, \nu, x) \Big[\mathrm{IsFreeAt}(\nu, n, x)$$

$$\wedge \neg \Big(\exists p < \mathrm{freepos}(k, \nu, x) \Big[n < p \wedge \mathrm{IsFreeAt}(\nu, p, x) \Big] \Big) \Big]$$

"Why does it count backwards?" Tetra asked.
"You'll see," Miruka said.

Definition 29. $\mathrm{freenum}(\nu, x)$ *is a fuction that returns the number of positions where* ν *is* $\langle\!\langle free \rangle\!\rangle$ *in* x.

$$\mathrm{freenum}(\nu, x) \overset{\mathrm{def}}{=} \min n \leqslant \mathrm{len}(x) \Big[\mathrm{freepos}(n, \nu, x) = 0 \Big]$$

Definition 30. $\mathrm{substSome}(k, x, \nu, c)$ *is a fuction that returns a* $\langle\!\langle formula \rangle\!\rangle$ *in which* k *instances of* ν *where* ν *is free in* x *are replaced by* c.

$$\begin{cases} \mathrm{substSome}(0, x, \nu, c) & \overset{\mathrm{def}}{=} x \\ \mathrm{substSome}(k + 1, x, \nu, c) & \overset{\mathrm{def}}{=} \quad \mathrm{substAtWith}(\mathrm{substSome}(k, x, \nu, c), \\ & \qquad\qquad\qquad\qquad \mathrm{freepos}(k, \nu, x), c) \end{cases}$$

"Oh, I get it!" I said.
"You get what?" Tetra asked.
"Why $\mathrm{freepos}(k, \nu, x)$ counts backwards. When you calculate $\mathrm{substSome}(k, x, \nu, c)$ the k decreases, so when you get to the end of the sequence it's down to 0 and it ends."

Definition 31. $\mathsf{subst}(a, v, c)$ *is a fuction that returns a* ⟨⟨*formula*⟩⟩ *in which all instances of* v *where* v *is free in* a *are replaced by* c.

$$\mathsf{subst}(a, v, c) \overset{\text{def}}{=} \mathsf{substSome}(\mathsf{freenum}(v, a), a, v, c)$$

"$\mathsf{subst}(a, v, c)$ here replaces all free instances of v in the a that showed up in axiom III-1[6] with c," Miruka said.

Definition 32. $\mathsf{implies}(a, b)$, $\mathsf{and}(a, b)$, $\mathsf{equiv}(a, b)$, $\mathsf{exists}(x, a)$ *are respectively functions that return* ⟨⟨(a) $\Rightarrow$ (b)⟩⟩, ⟨⟨(a) $\wedge$ (b)⟩⟩, ⟨⟨(a) $\Leftrightarrow$ (b)⟩⟩, *and* ⟨⟨$\exists x(a)$⟩⟩.[a]

$$\mathsf{implies}(a, b) \overset{\text{def}}{=} \mathsf{or}(\mathsf{not}(a), b)$$

$$\mathsf{and}(a, b) \overset{\text{def}}{=} \mathsf{not}(\mathsf{or}(\mathsf{not}(a), \mathsf{not}(b)))$$

$$\mathsf{equiv}(a, b) \overset{\text{def}}{=} \mathsf{and}(\mathsf{implies}(a, b), \mathsf{implies}(b, a))$$

$$\mathsf{exists}(x, a) \overset{\text{def}}{=} \mathsf{not}(\mathsf{forall}(x, \mathsf{not}(a)))$$

---

[a]see p. 293

Definition 33. $\mathsf{typelift}(n, x)$ *is a fuction that returns* x ⟨⟨*typelifted*⟩⟩ *by* n. *The part where* $\mathsf{prime}(1, x[k])^n$ *is being multiplied is the typelift. Note that the function distinguishes between variables and constants, and the latter are not altered.*

$$\mathsf{typelift}(n, x) \overset{\text{def}}{=} \min y \leqslant x^{(x^n)} \left[\forall k \leqslant \mathsf{len}(x) \left[\right. \right.$$

$$\left(\neg \mathsf{IsVar}(x[k]) \wedge y[k] = x[k] \right)$$

$$\left. \left. \vee \left(\mathsf{IsVar}(x[k]) \wedge y[k] = x[k] \times \mathsf{prime}(1, x[k])^n \right) \right] \right]$$

"What's it mean to typelift something?" Yuri asked.

---

[6]see p. 295

"You just increase the type of each variable in the sequence by one," Miruka said. "Say for example that x is $x_2(x_1)$. Then as a symbol sequence you have $\boxed{x_2}\,\boxed{(}\,\boxed{x_1}\,\boxed{)}$. Then $\texttt{typelift}(1, x)$ would be $\boxed{x_3}\,\boxed{(}\,\boxed{x_2}\,\boxed{)}$, and $\texttt{typelift}(2, x)$ is $\boxed{x_4}\,\boxed{(}\,\boxed{x_3}\,\boxed{)}$."

"So nothing happens to the constants $\boxed{(}$ and $\boxed{)}$, just to $\boxed{x_2}$ and $\boxed{x_1}$... Got it!"

"More programming-style math!" Tetra said.

"Yeah?" I said.

"Sure, because it's splitting processing into cases according to $\texttt{IsVar}(x[k])$."

"Also, it's a loop for $\forall\, k \leqslant \texttt{len}(x)$ with $\texttt{len}(x)$ as the terminating condition," Miruka said.

"And to think Gödel did all this in an era with no computers... Amazing!" Tetra said.

10.8.5 Axioms, Theorems, and Formal Proofs

"Time to set up our axioms," Miruka said.

Definition 34. $\texttt{IsAxiomI}(x)$ *is a predicate meaning 'x is a $\langle\!\langle$propositional formula$\rangle\!\rangle$ derived from axiom I[a].' Say that the Gödel numbers for axioms I-1, I-2, and I-3 are* $\alpha_1, \alpha_2, \alpha_3,$ *respectively. Then,*

$$\texttt{IsAxiomI}(x) \overset{\text{def}}{\Longleftrightarrow} x = \alpha_1 \lor x = \alpha_2 \lor x = \alpha_3.$$

---

[a]see p. 294

"Why the smile?" I asked.

"Because we're finally getting to where we can write formal proofs," Miruka said.

Definition 35. IsSchemaII(n, x) *is a predicate meaning 'x is a $\langle\!\langle propositional\ formula\rangle\!\rangle$ derived from axiom II-n^a.'*

$$\text{IsSchemaII}(1, x) \overset{\text{def}}{\iff} \exists\, p \leqslant x \left[\text{IsForm}(p) \right.$$
$$\left. \wedge\ x = \text{implies}(\text{or}(p, p), p) \right]$$
$$\text{IsSchemaII}(2, x) \overset{\text{def}}{\iff} \exists\, p, q \leqslant x \left[\text{IsForm}(p) \wedge \text{IsForm}(q) \right.$$
$$\left. \wedge\ x = \text{implies}(p, \text{or}(p, q)) \right]$$
$$\text{IsSchemaII}(3, x) \overset{\text{def}}{\iff} \exists\, p, q \leqslant x \left[\text{IsForm}(p) \wedge \text{IsForm}(q) \right.$$
$$\left. \wedge\ x = \text{implies}(\text{or}(p, q), \text{or}(q, p)) \right]$$
$$\text{IsSchemaII}(4, x) \overset{\text{def}}{\iff} \exists\, p, q, r \leqslant x \left[\text{IsForm}(p) \wedge \text{IsForm}(q) \wedge \text{IsForm}(r) \right.$$
$$\left. \wedge\ x = \text{implies}(\text{implies}(p, q), \text{implies}(\text{or}(r, p), \text{or}(r, q))) \right]$$

^{a}see p. 295

Definition 36. IsAxiomII(x) *is a predicate meaning 'x is a $\langle\!\langle propositional\ formula\rangle\!\rangle$ derived from axiom II^a.'*

$$\text{IsAxiomII}(x) \overset{\text{def}}{\iff}$$
$$\text{IsSchemaII}(1, x) \vee \text{IsSchemaII}(2, x) \vee$$
$$\text{IsSchemaII}(3, x) \vee \text{IsSchemaII}(4, x)$$

^{a}see p. 295

Definition 37. IsNotBoundIn(z, y, v) *is a predicate meaning 'z does not contain a $\langle\!\langle variable\rangle\!\rangle$ that is $\langle\!\langle bound\rangle\!\rangle$ anywhere in y where v is $\langle\!\langle free\rangle\!\rangle$.'*

$$\text{IsNotBoundIn}(z, y, v) \overset{\text{def}}{\iff}$$
$$\neg \Big(\exists\, n \leqslant \text{len}(y) \left[\exists\, m \leqslant \text{len}(z) \left[\exists\, w \leqslant z \right.\right.$$
$$\left[w = z[m] \wedge \text{IsBoundAt}(w, n, y) \wedge \right.$$
$$\left.\left.\left. \text{IsFreeAt}(v, n, y) \right] \right] \right] \Big)$$

Definition 38. $\mathtt{IsSchemaIII}(1, x)$ *is a predicate meaning* 'x *is a $\langle\!\langle$propositional formula$\rangle\!\rangle$ derived from axiom III-1[a].*'

$$\mathtt{IsSchemaIII}(1, x) \overset{\text{def}}{\iff}$$

$$\exists\, v, y, z, n \leqslant x \left[\mathtt{IsVarType}(v, n) \wedge \mathtt{IsNthType}(z, n) \right.$$

$$\wedge\ \mathtt{IsForm}(y) \wedge \mathtt{IsNotBoundIn}(z, y, v)$$

$$\left. \wedge\ x = \mathtt{implies}(\mathtt{forall}(v, y), \mathtt{subst}(y, v, z)) \right.$$

---

[a]see p. 295

Definition 39. $\mathtt{IsSchemaIII}(2, x)$ *is a predicate meaning* 'x *is a $\langle\!\langle$propositional formula$\rangle\!\rangle$ derived from axiom III-2[a].*'

$$\mathtt{IsSchemaIII}(2, x) \overset{\text{def}}{\iff}$$

$$\exists\, v, q, p \leqslant x \left[\mathtt{IsVar}(v) \wedge \mathtt{IsForm}(p) \wedge \right.$$

$$\neg\mathtt{IsFree}(v, p) \wedge \mathtt{IsForm}(q) \wedge$$

$$\left. x = \mathtt{implies}(\mathtt{forall}(v, \mathtt{or}(p, q)), \mathtt{or}(p, \mathtt{forall}(v, q))) \right]$$

---

[a]see p. 296

Definition 40. $\mathtt{IsAxiomIV}(x)$ *is a predicate meaning* 'x *is a $\langle\!\langle$propositional formula$\rangle\!\rangle$ derived from axiom IV[a].*'

$$\mathtt{IsAxiomIV}(x) \overset{\text{def}}{\iff}$$

$$\exists\, u, v, y, n \leqslant x \left[\mathtt{IsVarType}(u, n + 1) \wedge \mathtt{IsVarType}(v, n) \right.$$

$$\wedge\ \neg\mathtt{IsFree}(u, y) \wedge \mathtt{IsForm}(y)$$

$$\left. \wedge\ x = \mathtt{exists}(u, \mathtt{forall}(v, \mathtt{equiv}(\langle u \rangle * \mathtt{paren}(\langle v \rangle), y))) \right]$$

---

[a]see p. 296

Definition 41. IsAxiomV(x) *is a predicate meaning 'x is a $\langle\!\langle$propositional formula$\rangle\!\rangle$ derived from axiom V[a].' Say that the Gödel number for axiom V is α_4. Then,*

$$\text{IsAxiomV}(x) \overset{\text{def}}{\Longleftrightarrow} \exists\, n \leqslant x \left[x = \text{typelift}(n, \alpha_4) \right].$$

[a]see p. 296

Definition 42. IsAxiom(x) *is a predicate meaning 'x is an $\langle\!\langle$axiom$\rangle\!\rangle$.'*

$$\text{IsAxiom}(x) \overset{\text{def}}{\Longleftrightarrow}$$

$$\text{IsAxiomI}(x) \vee \text{IsAxiomII}(x) \vee$$

$$\text{IsAxiomIII}(x) \vee \text{IsAxiomIV}(x) \vee \text{IsAxiomV}(x)$$

where IsAxiomIII(x) *is defined as*

$$\text{IsAxiomIII}(x) \overset{\text{def}}{\Longleftrightarrow} \text{IsSchemaIII}(1, x) \vee \text{IsSchemaIII}(2, x).$$

"Now we need an inference rule," Miruka said.

Definition 43. IsConseq(x, a, b) *is a predicate meaning 'x is an $\langle\!\langle$immediate consequence$\rangle\!\rangle$ of a and b.'*

$$\text{IsConseq}(x, a, b) \overset{\text{def}}{\Longleftrightarrow}$$

$$a = \text{implies}(b, x) \vee \exists\, v \leqslant x \left[\text{IsVar}(v) \wedge x = \text{forall}(v, a) \right]$$

"The part before the $\vee$ means $a = \text{implies}(b, x)$ and we're obtaining x from b, right?" Tetra said.

"It's equivalent to $b \rightarrow x$ and obtaining x from b, yes," Miruka said.

"And the part after the $\vee$ means we're getting $\text{forall}(v, a)$ from a?"

"It's equivalent to getting $\forall v(a)$ from a, yes."

Definition 44. IsProof(x) *is a predicate meaning 'x is a ⟨⟨formal proof⟩⟩.'*

$$\text{IsProof}(x) \stackrel{\text{def}}{\Longleftrightarrow} \text{len}(x) > 0$$
$$\land\, \forall n \leqslant \text{len}(x) \left[n > 0 \Rightarrow \text{IsAxiomAt}(x, n) \lor \text{ConseqAt}(x, n) \right.$$

where IsAxiomAt(x, n) *and* ConseqAt(x, n) *are respectively defined as*

$$\text{IsAxiomAt}(x, n) \stackrel{\text{def}}{\Longleftrightarrow} \text{IsAxiom}(x[n]),$$
$$\text{ConseqAt}(x, n) \stackrel{\text{def}}{\Longleftrightarrow} \exists p, q < n \left[p, q > 0 \land \text{IsConseq}(x[n], x[p], x \right.$$

Definition 45. Proves(p, x) *is a predicate meaning 'p is a ⟨⟨formal proof⟩⟩ for x.'*

$$\text{Proves}(p, x) \stackrel{\text{def}}{\Longleftrightarrow} \text{IsProof}(p) \land \text{IsEndedWith}(p, x)^a$$

---

[a]see Def. 23 on p. 324 for the definition of the IsEndedWith(p, x) function.

"Yuri," Miruka said.

"Yes?"

"What's going on in this last one here?"

"We have a p that proves x!"

"Winter is thawing out at last, isn't it?" Tetra said.

Definition 46. IsProvable(x) *is a predicate meaning 'A ⟨⟨formal proof⟩⟩ of x exists.'*

$$\text{IsProvable}(x) \stackrel{\text{def}}{\Longleftrightarrow} \exists p \left[\text{Proves}(p, x) \right]$$

Miruka smiled.

"Time for a quiz," she said. "What's the big difference between definitions 1–45 and 46?"

We all fell silent.

"Is it that definition 46 only has one free variable?" Tetra asked.

"No. We've defined plenty of predicates like that."

"Is it that the form is different?" Yuri asked.

"Details."

"You know, the $\exists p$ here."

"We're using $\exists$ all over the place, aren't we?" Miruka said, her eyes twinkling.

"Yeah, but always for something like $\exists p \leqslant M$."

"You've got it," Miruka said. "In definitions 1 through 45, there's an upper limit every time we use a $\forall$ or an $\exists$. If you think of all this as a system for repeating $\forall$ and $\exists$ operations to find propositions, the upper limit on those tell you how many times you'll be repeating—that's what makes them primitive recursive. So all of the functions and predicates in definitions 1 through 45 are primitive recursive. Only IsProvable(x) in definition 46 here isn't."

10.9   THE NEW SPRING: UNDECIDABLE STATEMENTS

10.9.1   A Change of Seasons

"We're heading into the New Spring," Miruka said. "But before that, let's review the seasons we've passed through."

She popped another chocolate into her mouth, closing her eyes and smiling for a moment before continuing.

"In Spring, we defined our formal system P. Specifically, we decided on the elementary symbols, axioms, and inference rules we'll use.

"In Summer, we defined Gödel numbers and how we would associate them with the elementary symbols and sequences in our formal system. Doing this allowed us to represent everything in the formal system as a number.

"In Fall, we defined primitive recursive functions and predicates. We also learned about the representation theorem, though we haven't proved that one yet. The representation theorem serves as our bridge between the world of meaning and the world of formalism.

"In Winter, we defined $\text{Proves}(p, x)$, a primitive recursive predicate that says p is a formal proof of x."

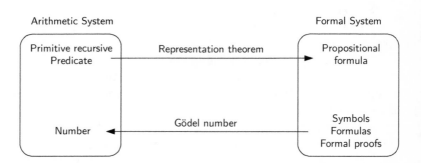

"And now...the New Spring. We're going to use the previous year of preparation to construct decidable statements. Somebody tell me why we want to do that."

"To show that our formal system P contains *un*decidable statements," I said. "Statements for which we can prove neither A nor ¬A."

Miruka nodded.

"There are eight parts to the New Spring," she said. "We're going to plant some seeds, watch them sprout and grow branches, leaves, and buds, and then we'll harvest plums and peaches. Then we can sit back and admire the cherry blossoms."

"Cherry blossoms?" I said.

"The completed proof of the first incompleteness theorem."

10.9.2   The Seeds: From Meaning to Formalism

"We start by defining this bivariate predicate Q," Miruka said.

$$Q(x, y) \overset{\text{def}}{\iff} \neg\text{Proves}(x, \text{subst}(y, \boxed{y_1}, \overline{y}))$$

"This predicate $Q(x, y)$ says that x is not a formal proof of $\text{subst}(y, \boxed{y_1}, \overline{y})$."

Miruka turned to Tetra.

"Is $Q(x, y)$ a primitive recursive predicate?" she asked.

"Um ... yes, it should be," Tetra said.

"Why?"

"Because it only uses primitive recursive predicates and functions that we defined in Winter."

"Good." Miruka nodded. "To make things easier to read, we'll define a couple more Gödel numbers, like this."

$$\boxed{x_1} \stackrel{\text{def}}{=} 17, \quad \boxed{y_1} \stackrel{\text{def}}{=} 19$$

"Hang on," Yuri said. "What is it we defined in Winter?"

"$\mathsf{Proves}(p, x)$ and $\mathsf{subst}(x, \nu, c)$," I said.

"Don't forget the function $\overline{x}$ for obtaining numerals!" added Tetra.

"And what's $\boxed{y_1}$?"

"Just a number, 19."

"And why does $\boxed{y_1}$ get to be 19?"

"From the definition of Gödel numbers[7]."

"And since it's just a number it's a constant function, which we said makes it primitive recursive," Miruka said.

Yuri nodded. "I'm good now," she said.

"Okay, follow this closely," Miruka said. "Now we can apply FALL-2 from the representation theorem[8] and see that there must exist some bivariate formula q that makes this true for all numbers m and n."

$$\neg Q(m, n) \Rightarrow \text{there exists a } \langle\!\langle \text{formal proof} \rangle\!\rangle \text{ for } \mathsf{not}(q\langle \overline{m}, \overline{n} \rangle)$$

"Here, we're defining $q\langle \overline{m}, \overline{n} \rangle$ like this."

$$q\langle \overline{m}, \overline{n} \rangle \stackrel{\text{def}}{=} \mathsf{subst}(\mathsf{subst}(q, \boxed{x_1}, \overline{m}), \boxed{y_1}, \overline{n})$$

"Wouldn't this mean you're using q to define q?" Tetra asked.

"No. We're using q to define $q\langle \overline{m}, \overline{n} \rangle$."

"I'm sorry, I guess I'm not quite seeing the difference."

---

[7]see p. 301

[8]see p. 309

"Read it carefully—q is the Gödel number for a bivariate formula. The Gödel numbers for the variables are $\boxed{x_1}$ and $\boxed{y_1}$."

"Which are 17 and 19, right."

"So that means $q\langle \overline{m}, \overline{n} \rangle$ is the Gödel number for a statement in which the two variables in q are replaced by $\overline{m}$ and $\overline{n}$."

"Oh, I get it! We did something similar when we talked about the representation theorem, didn't we."

Miruka nodded. "To make the explanation easier I've moved on to what we get from FALL-2, but we have to treat FALL-1 and FALL-2 together. In other words, the q here is the same thing in FALL-1 and FALL-2. We'll get back to FALL-1 in a minute. But for now we represent 'there exists a formal proof for $not(q\langle \overline{m}, \overline{n} \rangle)$' as $IsProvable(not(q\langle \overline{m}, \overline{n} \rangle))$, which gives us this A0."

▶ A0

$$\neg Q(m, n) \Rightarrow \underset{\wwwww}{IsProvable(not(q\langle \overline{m}, \overline{n} \rangle))}$$

"From the definition of predicate Q,[9] we can write $\neg Q(m, n)$ as $\neg \neg Proves(m, n\langle \overline{n} \rangle)$, in other words $Proves(m, n\langle \overline{n} \rangle)$. Here, we're defining $n\langle \overline{n} \rangle$ like this."

$$n\langle \overline{n} \rangle \overset{\text{def}}{=} subst(n, \boxed{y_1}, \overline{n})$$

"Then we can get this A1 from A0."

▶ A1

$$\underset{\wwwww}{Proves(m, n\langle \overline{n} \rangle)} \Rightarrow IsProvable(not(q\langle \overline{m}, \overline{n} \rangle))$$

"We'll use A1 here when we look at the 'leaves' stage."

"What exactly is $subst(n, \boxed{y_1}, \overline{n})$?" Tetra asked.

"I guess that would be a statement where the free variable $\boxed{y_1}$ in the formula n is replace by, uh . . . $\overline{n}$, the numeral form of n," I said.

"Right," Miruka said. "From a metamathematical perspective, that's a good way to put it. From an arithmetic perspective, things get messy. The formulas all become Gödel numbers, the numeralized

---

[9]see p. 334

$\overline{n}$ becomes the Gödel number for $\overline{n}$ ... Gödel numbers everywhere. *Everything* ends up a number."

"Um, so what exactly *is* $\mathtt{subst}(n, \boxed{y_1}, \overline{n})$?" Tetra asked.

"The diagonalization of n," Miruka said. "In other words, if you think of n as saying '*something* is so-and-so,' then $\mathtt{subst}(n, \boxed{y_1}, \overline{n})$ is saying '*something* is so-and-so is so-and-so.'"

"Ha! Diagonalization rears its head yet again!" I said.

"Let's get back to FALL-1 of the representation theorem[10]," Miruka said. "We need it to find B0 here."

▶ B0

$\qquad Q(m, n) \Rightarrow \mathtt{IsProvable}(q \langle \overline{m}, \overline{n} \rangle)$

"From the definition of predicate Q,[11] we can write $Q(m, n)$ as $\neg \mathtt{Proves}(m, n \langle \overline{n} \rangle)$. Then we can use B0 to get B1."

▶ B1

$\qquad \neg\mathtt{Proves}(m, n \langle \overline{n} \rangle) \Rightarrow \mathtt{IsProvable}(q \langle \overline{m}, \overline{n} \rangle)$

"We'll use B1 when we talk about buds."

Tetra was flipping back and forth in her notes.

"Looking for something?" I asked.

"Don't mind me," she said. "Just checking on something."

10.9.3   The Sprouts: Defining p

"There are two free variables in q," Miruka said, "so we'll let those be $\boxed{x_1}$ and $\boxed{y_1}$, and write q like this."

$$q = q \langle \boxed{x_1}, \boxed{y_1} \rangle$$

"Now if we define the formula p as $\mathtt{forall}(\boxed{x_1}, q)$, we can write p like this."

$$p \overset{\text{def}}{=} \mathtt{forall}(\boxed{x_1}, q \langle \boxed{x_1}, \boxed{y_1} \rangle)$$

"If you look closely at this, you'll see that the free variable $\boxed{x_1}$ in q is bound by $\mathtt{forall}(\boxed{x_1}, \cdots)$ in p, so as you can see $\boxed{y_1}$ is the only free variable in p. Let's write p as $p \langle \boxed{y_1} \rangle$. Then we can write C1 like this."

---

[10] see p. 309
[11] see p. 334

▶ C1

$$p\langle\boxed{y_1}\rangle = \texttt{forall}(\boxed{x_1}, \texttt{q}\langle\boxed{x_1}, \boxed{y_1}\rangle)$$

"Then we can replace $\boxed{y_1}$ with $\overline{p}$, giving us C2 like this."

▶ C2

$$p\langle\overline{p}\rangle = \texttt{forall}(\boxed{x_1}, \texttt{q}\langle\boxed{x_1}, \overline{p}\rangle)$$

"Now we can say this."

$$p\langle\overline{p}\rangle \overset{\text{def}}{=} \texttt{subst}(p, \boxed{y_1}, \overline{p})$$

"We'll use C2 here when we talk about leaves."

10.9.4   The Branches: Defining r

"In the branches stage," Miruka said, "we define the univariate formula r as $\texttt{q}\langle\boxed{x_1}, \overline{p}\rangle$. The remaining free variable in r is $\boxed{x_1}$, so we can write r as $r\langle\boxed{x_1}\rangle$."

▶ C3

$$r\langle\boxed{x_1}\rangle \overset{\text{def}}{=} \texttt{q}\langle\boxed{x_1}, \overline{p}\rangle$$

"Now we replace the $\boxed{x_1}$ in C3 with $\overline{m}$, which gives us C4."

▶ C4

$$r\langle\overline{m}\rangle = \texttt{q}\langle\overline{m}, \overline{p}\rangle$$

"C4 is another one we'll use when we talk about leaves."

Miruka paused and looked off in the distance for a moment.

"One thing I don't want you to forget," she said. "We're still in the world of meaning. From an arithmetic perspective we're just using numbers, but they're Gödel numbers. So from a metamathematical perspective it's all numerals, formulas, and formal proofs."

10.9.5 The Leaves: The Flow From A1

"On to the leaves then," Miruka said. "Here we want to use the A1 we defined in the seeds stage[12] to represent r."

▶ A1

$\text{Proves}(m, n\langle\overline{n}\rangle) \Rightarrow \text{IsProvable}(\text{not}(q\langle\overline{m}, \overline{n}\rangle))$

"First, we substitute the n's in A1 with p's, and get A2."

▶ A2

$\text{Proves}(m, p\langle\overline{p}\rangle) \Rightarrow \text{IsProvable}(\text{not}(q\langle\overline{m}, \overline{p}\rangle))$

"Now we're going to combine a bunch of As with the Cs we derived before. First, combine A2 and C2 ($p\langle\overline{p}\rangle = \text{forall}(\boxed{x_1}, q\langle\boxed{x_1}, \overline{p}\rangle)$) to get A3."

▶ A3

$\text{Proves}(m, \underline{\text{forall}(\boxed{x_1}, q\langle\boxed{x_1}, \overline{p}\rangle)}) \Rightarrow \text{IsProvable}(\text{not}(q\langle\overline{m}, \overline{p}\rangle))$

"Next, combine A3 and C3 ($r\langle\boxed{x_1}\rangle = q\langle\boxed{x_1}, \overline{p}\rangle$) to get A4."

▶ A4

$\text{Proves}(m, \text{forall}(\boxed{x_1}, \underline{r\langle\boxed{x_1}\rangle})) \Rightarrow \text{IsProvable}(\text{not}(q\langle\overline{m}, \overline{p}\rangle))$

"Finally, combine A4 and C4 ($r\langle\overline{m}\rangle = q\langle\overline{m}, \overline{p}\rangle$) to get A5."

▶ A5

$\text{Proves}(m, \text{forall}(\boxed{x_1}, r\langle\boxed{x_1}\rangle)) \Rightarrow \text{IsProvable}(\text{not}(\underline{r\langle\overline{m}\rangle}))$

"We'll use A5 here when we talk about plums."

---

[12]see p. 336

10.9.6   The Buds: The Flow from B1

"Our goal in the buds stage," Miruka said, "is to use the B1 we derived in the seeds stage[13] to represent r."

▶ B1

$$\neg \texttt{Proves}(\mathfrak{m}, \mathfrak{n} \langle \overline{\mathfrak{n}} \rangle) \Rightarrow \texttt{IsProvable}(\mathfrak{q} \langle \overline{\mathfrak{m}}, \overline{\mathfrak{n}} \rangle)$$

"We can substitute the n's in B1 with p's, and get this B2."

▶ B2

$$\neg \texttt{Proves}(\mathfrak{m}, \underline{\mathfrak{p} \langle \overline{\mathfrak{p}} \rangle}) \Rightarrow \texttt{IsProvable}(\mathfrak{q} \langle \overline{\mathfrak{m}}, \overline{\mathfrak{p}} \rangle)$$

"Then we combine B2 and C2 $(\mathfrak{p} \langle \overline{\mathfrak{p}} \rangle = \texttt{forall}(\boxed{x_1}, \mathfrak{q} \langle \boxed{x_1}, \overline{\mathfrak{p}} \rangle))$ to get B3."

▶ B3

$$\neg \texttt{Proves}(\mathfrak{m}, \underline{\texttt{forall}(\boxed{x_1}, \mathfrak{q} \langle \boxed{x_1}, \overline{\mathfrak{p}} \rangle)}) \Rightarrow \texttt{IsProvable}(\mathfrak{q} \langle \overline{\mathfrak{m}}, \overline{\mathfrak{p}} \rangle)$$

"B3 and C3 $(\mathfrak{r} \langle \boxed{x_1} \rangle \overset{\texttt{def}}{=} \mathfrak{q} \langle \boxed{x_1}, \overline{\mathfrak{p}} \rangle)$ together give us B4..."

▶ B4

$$\neg \texttt{Proves}(\mathfrak{m}, \texttt{forall}(\boxed{x_1}, \mathfrak{r} \langle \boxed{x_1} \rangle)) \Rightarrow \texttt{IsProvable}(\mathfrak{q} \langle \overline{\mathfrak{m}}, \overline{\mathfrak{p}} \rangle)$$

"...and B4 and C4 $(\mathfrak{r} \langle \overline{\mathfrak{m}} \rangle = \mathfrak{q} \langle \overline{\mathfrak{m}}, \overline{\mathfrak{p}} \rangle)$ together give us B5."

▶ B5

$$\neg \texttt{Proves}(\mathfrak{m}, \texttt{forall}(\boxed{x_1}, \mathfrak{r} \langle \boxed{x_1} \rangle)) \Rightarrow \texttt{IsProvable}(\mathfrak{r} \langle \overline{\mathfrak{m}} \rangle)$$

"We'll use B5 when we talk about peaches."

---

[13]see p. 337

10.9.7 An Undecidable Statement

"Actually we've already produced a definition for undecidable statements," Miruka said, "in the buds stage."

$$\mathtt{forall}(\boxed{x_1}, r\langle\!\langle\boxed{x_1}\rangle\!\rangle)$$

"Let's call this statement g."

▷ **Definition of** g

$$g \overset{\text{def}}{=} \mathtt{forall}(\boxed{x_1}, r\langle\!\langle\boxed{x_1}\rangle\!\rangle)$$

"We need to give two proofs to show that g is an undecidable statement."

· ¬IsProvable(g)

· ¬IsProvable(not(g))

"These two proofs will be the plums and peaches that we harvest."

10.9.8 The Plums: Proof of ¬IsProvable(g)

"Start by assuming that our formal system P is consistent," Miruka said. "In other words, that it contains no contradictions."

▶ D0

Formal system P is consistent

"The proposition we want to prove is ¬IsProvable($\mathtt{forall}(\boxed{x_1}, r\langle\!\langle\boxed{x_1}\rangle\!\rangle)$). We're going to use proof by contradiction, so let D1 be the negation of the proposition we want to prove."

▶ D1

IsProvable($\mathtt{forall}(\boxed{x_1}, r\langle\!\langle\boxed{x_1}\rangle\!\rangle)$)

"Let the formal proof of $\mathtt{forall}(\boxed{x_1}, r\langle\!\langle\boxed{x_1}\rangle\!\rangle)$ in D1 be s. Then we get this D2."

▶ D2

Proves(s, $\mathtt{forall}(\boxed{x_1}, r\langle\!\langle\boxed{x_1}\rangle\!\rangle)$)

"Okay, let's go back and reread A5, which we got back in the leaves stage[14]. It says that A5 holds for all values of m, so it should

---

[14]see p. 339

also hold when we let m be the s of D2. That gives us this D3."

▶ D3

    $\text{Proves}(s, \text{forall}(\boxed{x_1}, r\langle\!\langle\boxed{x_1}\rangle\!\rangle)) \Rightarrow \text{IsProvable}(\text{not}(r\langle\bar{s}\rangle))$

"So we can use D2 and D3 to get this D4."

▶ D4

    $\text{IsProvable}(\text{not}(r\langle\bar{s}\rangle))$

"Now let's take a look at D1 and think about what it's saying about this formal system from a metamathematical viewpoint. We know there's a formal proof for the statement $\text{forall}(\boxed{x_1}, r\langle\!\langle\boxed{x_1}\rangle\!\rangle)$ in D1. In other words, it's a theorem. So let's combine that with axiom III-1,[15] which gives us $\text{subst}(r, \boxed{x_1}, \bar{s})$. In other words, $r\langle\bar{s}\rangle$. That means $r\langle\bar{s}\rangle$ is a theorem, so it too has a formal proof, giving us D5."

▶ D5

    $\text{IsProvable}(r\langle\bar{s}\rangle)$

"So D4 and D5 have told us that there are formal proofs for both $\text{not}(r\langle\bar{s}\rangle)$ and $r\langle\bar{s}\rangle$. That gives us D6."

▶ D6

    Formal system P is inconsistent

"And this is a contradiction with our premise D0, which said formal system P is *not* inconsistent. So by proof by contradiction, we've shown that the negation of D1, $\text{IsProvable}(\text{forall}(\boxed{x_1}, r\langle\!\langle\boxed{x_1}\rangle\!\rangle))$, holds. In other words, we've proven D7."

▶ D7

    $\neg\text{IsProvable}(\text{forall}(\boxed{x_1}, r\langle\!\langle\boxed{x_1}\rangle\!\rangle))$

---

[15]see p. 295

"Correct me if I'm wrong," Tetra said, "but didn't we just see two kinds of contradiction?"

"How clever of you," Miruka said.

"Two kinds?" I said. "How's that?"

"D4 and D5 are contradictions in the formal world," Miruka said. "They say that both a formula and its negation have a formal proof."

"Sure, I see that. So?"

"D0 and D6 are contradictions in the world of meaning. They say that both a proposition and its negation hold."

"Interesting! I'd never even considered that there might be different kinds of contradiction..."

10.9.9 The Peaches: Proof of ¬IsProvable(not(g))

"Time to harvest peaches," Miruka said. "First, we assume this E0."

▶ E0

    Formal system P is $\overset{\text{omega}}{\omega}$-consistent

"What's ω-consistent?" I asked.

"Something that needs a definition," Miruka said. "How about this?"

▷ **ω-inconsistency**

    A formal sytem is *ω-inconsistent* if for some univariate formula $f\langle\boxed{x_1}\rangle$ the following two conditions hold:

· There exists a formal proof for each of $f\langle\overline{0}\rangle, f\langle\overline{1}\rangle, f\langle\overline{2}\rangle, \ldots$.

· There exists a formal proof for $not(forall(\boxed{x_1}, f\langle\boxed{x_1}\rangle))$

▷ **ω-consistency**

    A formal sytem is *ω-consistent* if it is not ω-inconsistent.

"Omega consistency is a stricter condition than 'normal' consistency," Miruka said. "If a formal system is ω-consistent, then it is definitely consistent. But the reverse isn't necessarily the case—a formal system that's consistent isn't necessarily ω-consistent."

"That's...odd," I said. "You mean there's a formal proof of $f\langle t \rangle$ for all values of t, but there's also a formal proof for $\mathtt{not(forall(\boxed{x_1}, f\langle\!\langle \boxed{x_1} \rangle\!\rangle))}$?"

"Yeah, it sounds weird when you're using the normal interpretation of numbers. But making an assertion about all numbers and making one regarding ∀ are different things."

"Why the ω?" Tetra asked.

"It comes from the set of all numbers $\omega = \{0, 1, 2, \ldots\}$. The term ω-consistent means something is consistent using the standard interpretation of numbers."

"Got it," Tetra said, jotting this down in her notebook.

"Back to peaches," Miruka said. "Here's what we want to prove."

$$\neg\mathtt{IsProvable(not(forall(\boxed{x_1}, r\langle\!\langle \boxed{x_1} \rangle\!\rangle)))}$$

"The D7 we got when we harvested plums says that E1 holds for all sequences of formulas t."

▶ E1
$$\neg\mathtt{Proves(t, forall(\boxed{x_1}, r\langle\!\langle \boxed{x_1} \rangle\!\rangle))}$$

"In the buds stage we got B5,[16] which should hold when m is t, giving us E2."

▶ E2
$$\neg\mathtt{Proves(t, forall(\boxed{x_1}, r\langle\!\langle \boxed{x_1} \rangle\!\rangle))} \Rightarrow \mathtt{IsProvable(r\langle t \rangle)}$$

"From E1 and E2, E3 should hold for all values of t."

▶ E3
$$\mathtt{IsProvable(r\langle t \rangle)}$$

"Proof by contradiction again. E4 is the negation of the proposition we want to prove."

▶ E4
$$\mathtt{IsProvable(not(forall(\boxed{x_1}, r\langle\!\langle \boxed{x_1} \rangle\!\rangle)))}$$

---

[16]see p. 340

"From E4 and the fact that E3 holds for all values of t, we get E5."

▶ E5

Formal system P is ω-inconsistent

"But E5 contradicts our assumption E0, which says that formal system P is ω-consistent, so by proof by contradiction the negation of our assumption holds. In other words, we've proven E6."

▶ E6

$\neg\texttt{IsProvable(not(forall(}\boxed{x_1}\texttt{,}r\langle\boxed{x_1}\rangle\texttt{))))}$

10.9.10 The Cherry Blossoms: Proof that P is Incomplete

"And here we are," Miruka said. "From the D7 we harvested as plums, and E6 we harvested as peaches, we obtain F1."

▶ F1

There is no formal proof for either g or not(g)

"From F1, we get F2."

▶ F2

The formal system P is incomplete

"And with that, we've proven the first incompleteness theorem. Done and done. "

Yuri collapsed onto the desk with a groan.

"That was ... a good bit of work," I said.

Tetra was drawing what appeared to be a graph that took up a whole sheet in her notebook.

"What are you drawing there?" I asked.

"A map of our journey!" she said, holding up the page for us to see.

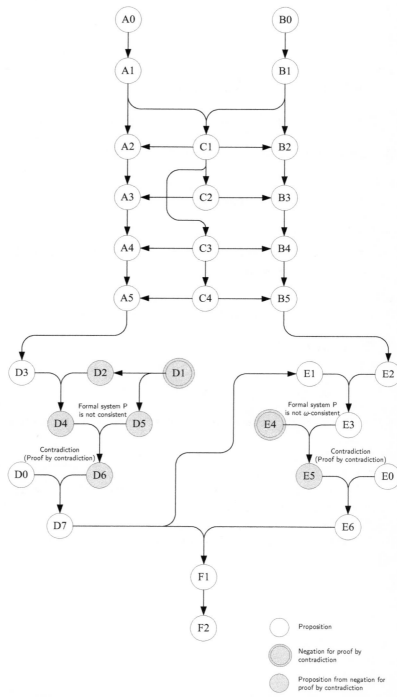

Outline of the proof of the first incompleteness theorem

10.10   WHAT THE INCOMPLETENESS THEOREMS MEAN

10.10.1   No Proof of Me Exists

I looked around the Chlorine room. The huge whiteboard was covered edge-to-edge in math. The table was buried under sheets of paper, the notes and diagrams that were the detritus of a long math session. I could almost feel all the definitions and axioms and formal proofs still ricocheting around inside my skull.

"Quite an impressive proof," I said.

"What we did was basically walk through Gödel's paper," Miruka said.

"Walked through a paper?" Yuri said. "Feels more like I've been raked over the coals."

"I love your diagram, by the way," I said to Tetra. "It's a nice map of where we've been."

Tetra twisted her mouth.

"I'm still not sure I quite get it all though," she said. "I can follow along with what we did in each season, up until the end. This undecidable statement we came up with in the New Spring part . . . I'm not sure exactly what that is."

"This statement g, right?" I said.

$$g = \texttt{forall}(\boxed{x_1}, r\langle\boxed{x_1}\rangle)$$

"If you want to think about what it means, it's better to write g in terms of p," Miruka said.

$$
\begin{aligned}
g &= \texttt{forall}(\boxed{x_1}, r\langle\boxed{x_1}\rangle) \\
&= \texttt{forall}(\boxed{x_1}, q\langle\boxed{x_1}, \overline{p}\rangle) && \text{from C3} \\
&= p\langle\overline{p}\rangle && \text{from C2}
\end{aligned}
$$

"Just like we defined in C1, p is a formula with one free variable, $\boxed{y_1}$. And g is $p\langle\overline{p}\rangle$, the diagonalization of p."

$$
\begin{aligned}
p &= p\langle\boxed{y_1}\rangle &&= \texttt{forall}(\boxed{x_1}, q\langle\boxed{x_1}, \boxed{y_1}\rangle) \\
g &= p\langle\overline{p}\rangle &&= \texttt{forall}(\boxed{x_1}, q\langle\boxed{x_1}, \overline{p}\rangle)
\end{aligned}
$$

"So that means g is what you get when you replace the $\boxed{y_1}$ in p with the numeral for p itself," I said.

Miruka nodded.

"So q says that x is not a formal proof of the diagonalization of y," she said, "and p says that no formal proof of the diagonalization of y exists. Then g says there is no formal proof for the diagonalization of 'there is no formal proof for the diagonalization of y.'"

"I can't follow that at all," Yuri said.

"Easier to see if I write it down," Miruka said, going to the whiteboard and erasing a section. "Look at it like this."

p says—

> there exists no formal proof of the diagonalization
> of y

g says—

> there is no formal proof of the diagonalization of
> 'there is no formal proof of the diagonalization of
> y'

"Put p in quotes, and stick it into its own y. That's the diagonalization of p. So g is just the diagonalization of p, and if you look carefully at what it says you can expand it out to write something like this."

Miruka added some lines to what she had written about g:

g says—

> there is no formal proof of the diagonalization of
> 'there is no formal proof of the diagonalization of
> y' $\longrightarrow$
>
> there is no formal proof of the diagonalization of
> p $\longrightarrow$
>
> there is no formal proof of g

"So in a metamathematical sense, the statement g is saying 'no proof of me exists.'"

"This is making me dizzy 'is making me dizzy,'" Yuri said.

"I'm still a bit dizzy myself, but I think I'm catching on," Tetra said. "If a formal proof of g did exist, then the very existence of such a proof would contradict what g itself is saying in a metamathematical sense."

"Bingo!" Miruka said, holding up a finger. "And Gödel's proof is in essence a very precise mathematical description of exactly what g is, what it asserts, and what it does not."

"Wow, did it take a lot of work to do that," Tetra said, thumbing through her notes.

"But very interesting in how he did it," Miruka said. "Not by treating formulas as formulas, but as numbers instead. That let him use numbers to create predicates in the formal system, and thereby make metamathematical assertions about it. He worked out how to make mathematical objects say something about themselves—self-reference, in other words. A lot of the proof is showing how primitive recursion and the representation theorem can demonstrate how the self-referential statement he created does in fact refer to itself. Also that it's not just this formal system P that's incomplete, but *any* formal system that allows self-reference."

"Self reference, huh?" Tetra said, adding this to her notebook.

"Write down, 'no proof of me exists,'" Miruka said.

"And 'I am a liar,'" Yuri suggested.

"How about 'I do not belong to myself,'" I said. "Because now I finally realize where I've seen something like this before—Russel's paradox, which says $x \notin x$."

Miruka nodded. "Remember the elementary formula $a(b)$ we created for the formal system P? There was a restriction that said a is of type-$\{n + 1\}$, while b is of type-n. In other words, we have to have type-$n \in$ type-$\{n + 1\}$. You'll never see something like Russell's paradox in our formal system P that says $x \notin x$, in other words $\neg(x \in x)$. You can't, because the types aren't right. The self-reference is avoided."

Miruka leaned forward and began speaking faster.

"However, that doesn't mean you're safe. Self-reference still rears its ugly head when you create a statement in which you've replaced the variables in a formula with the numeralization of the statement itself."

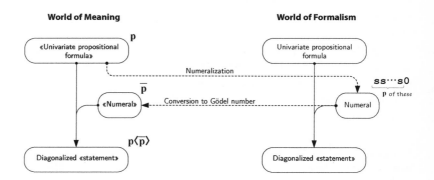

Self-reference through numeralization and Gödel numbers

"Ah!" Tetra said. "So that's why paying attention to those bracketed terms was so important."

"So Gödel has overcome Galileo's doubts," I said. "Russell was pretty slick, creating a self-referential paradox, but Gödel did him one better, using what looked like a bad situation and turning it into a proof."

"What do you mean, Galileo's doubts?" Yuri asked.

"That what looks like failure isn't necessarily so. That sometimes trudging on instead of turning back can lead you to unexpected worlds."

I stopped and remembered something my mother had said.

If there's no ground to walk on, spread your wings and fly.

10.10.2   Overview of the Second Incompleteness Theorem

"This convoluted statement g is an interesting one," Miruka said, "but it's an awfully . . . *artificial* thing to say."

> There exists no formal proof of the diagonaliza-
> tion of 'there exists no formal proof of the diago-
> nalization of g'

"But Gödel found another, more natural statement in this formal system that doesn't have a formal proof."

"Yeah? What's that?" I asked.

" 'I am consistent.' " Miruka said.

"That sounds strangely familiar..."

"It should—it's Gödel's second incompleteness theorem."

"No no no," Yuri said. "I can't take another proof just now."

"Just a quick-and-dirty overview then."

Yuri groaned, but Miruka continued on.

"The proof is for the proposition 'there exists no formal proof in formal system P for a statement representing a proposition stating that the formal system P is consistent.' Please recall our proposition D7[17] from the plum stage."

▶ D7
$$\neg\mathtt{IsProvable}(\mathtt{forall}(\boxed{x_1}, r\langle\!\langle\boxed{x_1}\rangle\!\rangle))$$

"The proof of D7 assumed D0, which says that the formal system P is consistent. Let's write the proposition 'Formal system P is consistent' as Consistent. From that, we can derive G1."

▶ G1
$$\underset{\sim}{\mathtt{Consistent}} \Rightarrow \neg\mathtt{IsProvable}(\mathtt{forall}(\boxed{x_1}, r\langle\!\langle\boxed{x_1}\rangle\!\rangle))$$

"Wait, I'm confused..." Tetra said.

Miruka paused. "Something wrong?"

"The plum stage was part of our proof of the *first* incompleteness theorem, right?"

"It was."

"Is it really okay to just, like, reach in and grab a line from the middle of another proof?"

"Ah, now that's the interesting part of the proof of the second incompleteness theorem." Miruka grinned.

"Specifically?"

"The first incompleteness theorem is a proof from a metamathematical perspective."

[17]see p. 343

"And that somehow allows us to define `Consistent`?" I asked.

"In Gödel's proof it was $\exists x \left[\texttt{IsForm}(x) \wedge \neg\texttt{IsProvable}(x) \right]$."

"Huh? So all you need is a formula that doesn't have a formal proof?"

"Yep. Because in a formal system that contains contradictions, you can give a formal proof of *any* formula."

Miruka grabbed a nearby sheet of paper and turned it to a blank side.

"We can combine G1 and C3 $(r\langle\!\langle\boxed{x_1}\rangle\!\rangle = q\langle\!\langle\boxed{x_1}\rangle, \overline{p}\rangle)$ to create G2."

▶ G2

    `Consistent` $\Rightarrow \neg\texttt{IsProvable}(\texttt{forall}(\boxed{x_1}, q\langle\!\langle\boxed{x_1}\rangle, \overline{p}\rangle)))$

"Next, from G2 and C2 $(p\langle\overline{p}\rangle = \texttt{forall}(\boxed{x_1}, q\langle\!\langle\boxed{x_1}\rangle, \overline{p}\rangle)))$ we get G3."

▶ G3

    `Consistent` $\Rightarrow \neg\texttt{IsProvable}(p\langle\overline{p}\rangle)$

"From G3, `Consistent` implies that no t can be a formal proof of $p\langle\overline{p}\rangle$. So that gives us G4."

▶ G4

    `Consistent` $\Rightarrow \forall t \left[\neg\texttt{Proves}(t, p\langle\overline{p}\rangle) \right]$

"Then we can use the predicate $Q(m, n)$ to write G4 as G5."

▶ G5

    `Consistent` $\Rightarrow \forall t \left[Q(t, p) \right]$

"Let c be a statement representing the proposition `Consistent`. Then a statement representing $\forall t \left[Q(t, p) \right]$ would be $\texttt{forall}(\boxed{x_1}, q\langle\!\langle\boxed{x_1}\rangle, \overline{p}\rangle)$. Then from G5, we get G6."

▶ G6

    $\texttt{IsProvable}(\texttt{implies}(c, \texttt{forall}(\boxed{x_1}, r\langle\!\langle\boxed{x_1}\rangle\!\rangle)))$

"Next, we assume G7 as the negation of ¬IsProvable(c), which is the proposition we want to prove."

▶ G7

    IsProvable(c)

"Applying our inference rule[18] to G6 and G7, we get G8."

▶ G8

    IsProvable(forall($\boxed{x_1}$, r($\boxed{x_1}$)))

"But G8 here contradicts D7, which we derived back in the plum phase. So by proof by contradiction the negation of G7 holds, giving us G9."

▶ G9

    ¬IsProvable(c)

"G9 says that there exists no formal proof of c. In other words, in formal system P there exists no formal proof of a formula representing the statement that formal system P is consistent. Done and done."

"Uh … wow," I said, blinking.

"To be fair, the parts where we derive G6 and G8 aren't totally obvious and are justifiably debatable," Miruka said. "But that would be straying beyond the bounds of Gödel's paper, and I'd say we've done enough today already."

10.10.3 The Fruits of the Incompleteness Theorems

We sat back for a breather, and to polish off the remaining snacks that were scattered about the table.

"Hey, Miruka," I said. "At the beginning of all this you said something about how the incompleteness theorems were actually a constructive contribution to mathematics. How are they used exactly?"

"I'm not a logician, so to be honest I couldn't give you a lot of details," Miruka said. "But I do know that, for example, you can use

---

[18]see p. 297

the second theorem to help uncover relationships between formal systems."

"Different formal systems can be related?" Tetra asked.

"Well, consider this," Miruka said. "Say you have some formal system X. Then you designate a formula a in X as a new axiom, and use that to define another formal system Y. Now that you've added an axiom, does Y have more theorems than X?"

We paused a moment to consider this.

"Well," I said, "it seems like if you have more axioms you should be able to give formal proofs of more formulas, so..."

"Not necessarily!" Yuri said. "For one thing, a might have already been a theorem in X, right?"

"That's exactly right," Miruka said, ruffling Yuri's hair. "If a already had a formal proof in X, then you haven't gained anything by calling it an axiom. Specifically, the set of all theorems in formal system X would equal the set of all theorems in Y."

"Sure, I'll buy that," I said.

"In fact," Miruka continued, "it's generally hard to determine whether you've really created a new formal system when you've done so by defining it on the basis of another one. *But*—"

Miruka leaned forward and raised her eyebrows.

"—but what if you find a way to use formal system Y to give a formal proof that formal system X is consistent?"

Silence again, but a briefer one.

"Ah, I get it!" I said.

"Me too!" Tetra said.

"Yeah, they're different, right?" Yuri said.

"In what way, Yuri?" Miruka asked.

"Because the second incompleteness theorem says you can't give a formal proof within a formal system of the system's own consistency. But if you can prove from within Y that X is consistent, then that means they can't really be the same system. Pretty cool!"

"Well done. So the second incompleteness theorem gives you a way to investigate the power of a formal system. In this case, if you can use Y to show that X is consistent, then in a sense Y is fundamentally more powerful than X."

"I second Yuri," I said. "That's pretty cool."

"And I third her," Tetra said. "I love the idea of taking an *inability* to prove something and turning that around into the *ability* to prove something. We've mathematically proven that not being able to do something isn't necessarily a weakness!"

10.10.4   The Limits of Mathematics?

We sat back pondering this for a while. After a time Yuri raised her hand.

"Question!" she said, then paused and bit her lip. "Er, maybe this is kind of a dumb thing to ask, but the incompleteness theorems didn't, like, punch a bunch of holes in mathematics, did they?"

"Not at all," Miruka said, then it was her turn to pause and think. "Well, honestly I guess it depends on your definition of 'a bunch of holes.' But it's not like they instantly invalidated every mathematical theorem that had ever been proven, and it's not like there's a bunch of neither-provable-nor-non-provable propositions fluttering about, getting in the way of mathematicians' research. I doubt most mathematicians even think much about them."

She leaned back and smiled.

"Don't let the dictionary meaning of the word 'incomplete' give you the wrong idea. These theorems are now part of the bedrock of modern logic. So it's not like they punched a bunch of holes that deflated mathematics—it's more like they punched a bunch of holes that let fresh air in."

"Well, it still makes me uneasy just the same," Tetra said, a serious expression on her face. "I used to think of mathematics as this absolute, solid thing. But now I know there are things out there that can neither be proven nor disproven, and that nothing can know if it contradicts itself without something else's help. So it still kinda feels like we've proven that even math has its limits, and I'm not sure I like that."

Miruka stood and wandered over to the window. She looked outside, toward the rapidly setting sun. After a few moments, she turned and faced Tetra.

"There's a problem with your argument," she said. "You need to define exactly what you mean when you use the word 'mathematics.' Do you mean 1," Miruka held up a single finger, "something where you can write down precise definitions and give some kind of proof of everything you can give a formal representation of? Or is it 2," a second finger went up, "something that isn't really definable, but exists in our hearts as something so grand as to deserve the name?"

Tetra looked down at the table.

"If it's the first definition you want to use," Miruka said, "then having examined the conditions of that thing called mathematics, we have to admit that it is subject to the incompleteness theorems and bound by what they say."

Tetra fidgeted in her seat.

Miruka continued. "But if you want to use the second definition, then the incompleteness theorems don't really apply, because you aren't really talking about math. You're talking about mathematical theory, or some kind of philosophy or aesthetics or something. Whatever it is, it isn't math."

Tetra looked up but remained silent.

Miruka looked at our faces in turn. "So here's what I think. If we want to talk about the incompleteness theorems themselves, then we need to limit our discussion to math, and math alone. If instead we're going to talk about mathy things like how you find the results of these theorems to be inspirational or depressing or whatever then that's fine too, but let's not fool ourselves into thinking that we're talking about actual math anymore."

"So you're saying that we can't represent mathematics as a formal system?" I said, smirking.

Miruka shook her head and chuckled.

"More like you can't use mathematics to give a definition of what mathematics is," she said. "Or that there's a difference between arguments within mathematics and ones about mathematics."

She pushed her glasses up her nose.

"But that's cool too. Seeing differences is another step toward understanding.

10.11 Riding on Dreams

10.11.1 Not the End

When we left the Narabikura Library it was already dark, more night than evening. The four of us followed the shrub-lined path that led down the hill, heading toward the train station. My mind was fuzzily wandering along the route from Leibniz's dream of computational logic, to Gödel's conversion of formal systems into numbers, to what I assumed led to the computers we have today.

We were tired from the long mathematical journey we'd taken that day, and I was glad to have seen it through to its end. Just as I was happy knowing that we'd go on other journeys, on other days.

"I hear waves," Miruka said. I stopped for a moment to listen carefully, and heard them myself, faint on the breeze.

When I started walking again I thought of rivers emptying into the ocean, their waters ending one journey and beginning another.

10.11.2 What is Mine

The train home was empty, just the four of us in a compartment, sitting across from each other in pairs. I sat across from Miruka. Yuri sat beside me across from Tetra. We were tired, but we were still buzzing from the intensity of today's session and couldn't resist talking about math and throwing little challenges at each other.

But eventually the rocking of the train got the better of us, and our words grew more sparse. Yuri let out a jaw-cracking yawn, and my eyelids soon grew heavy. Before I knew it, I had nodded off.

My eyes snapped open when the train shuddered through a curve. Tetra was asleep with her head on Miruka's shoulder, Yuri with hers on mine. Miruka was looking out the window, up at the night sky.

"Hey, Miruka—" I said.

She turned my way but shushed me and pointed at Yuri. I nodded my understanding.

Miruka held up a finger, then the same finger again, then two, then three. She cocked her head, waiting for my response. I held up my right hand, five fingers splayed wide. She rewarded me with a warm smile.

I recalled the spring day when we first met. Our first conversation, if you could call it that. Math problems amidst a storm of cherry blossoms. We'd had many more conversations since then, and I'd learned so much. The knowledge I'd gained was a treasure I would keep forever. But I was struck by a sudden thought, an instant certainty—it was a treasure that would be rendered worthless if I kept it to myself. Like Ay-Ay's music, like Tetra's language studies, it held value only when shared with others. Math and my love of learning would always be precious to me, but so would teaching.

It was almost April, the start of a new school year. There were new problems awaiting us, no doubt. A new journey was about to begin. But for now, it was a time to rest.

The train rattled on into the night. Tetra and Yuri dreamed beside us as Miruka and I gazed at each other in silent conversation.

I wished our ride would never end.

[Gödel's] idea was to use mathematical reasoning in exploring mathematical reasoning itself.

DOUGLAS HOFSTADTER
Gödel, Escher, Bach

Epilogue

"You got a minute?" the girl asked as she entered the math teacher's office. The wind that blew through the door she'd opened carried a hint of spring.

"Sure, what's up..." the teacher said. "Hey, are you okay? You don't look your usual chipper self."

"Nah, I'm good. Sharp as a tack, as usual."

"We'll see about that. Say you have three friends, Alice, Boris, and Chris, and you put either a red or a white hat on each."

"Oh, boy. Here we go..."

"All three can see the others' hats, but not their own."

"And they have to guess what color hat they're wearing, right?"

"Exactly. There are three red hats, and two white ones. After you give your friends their hats, you hide the others. Then you ask Alice what color her hat is, and she says she doesn't know. You following this?"

"Yeah, yeah," she said, waving a hand. "Keep going."

"Okay, so then you ask Boris what color his hat is, and he says he doesn't know."

"Red."

"Red what?"

"Chris's hat is red, right?"

"Er, I haven't finished the problem. I was supposed to tell you that Chris can see that Alice's and Boris's hats are red."

"Yep. And Chris's hat is still red, right?"

"Well, yeah, but how did you know when I hadn't even finished the problem?"

"Just think it through. Alice said she didn't know, so at least one of Boris or Chris must be red. Boris knows that too, so if he saw that Chris's hat was white, he'd know that his own must be red. But he didn't know for sure, so Chris's must be red."

"Impressive, as usual. Okay, you've bested me. So what brings you here today?"

"Um, I'm pretty sure I came here because I'd forgotten something, but now I've forgotten what I forgot."

"An interesting case of meta-forgetfulness. I'll bet it has something to do with graduation tomorrow."

"Maybe. A lot of my friends will be graduating, and I'm going to miss them. I hope I don't cry."

"Don't think of it as losing friends—think of it as seeing them off to the next stage in their journey."

"That's even worse somehow. Cut it out, or I'll start crying now."

"You should have seen the graduation ceremony we had in my junior year. It went down in school history as 'the Graduation of Tears.'"

The girl laughed.

"What's so funny?"

"The thought of you crying at graduation."

"It happens to the best of us."

"Oh, I remember now! You promised me a card!"

"Ah, right. Let's see..." The teacher fished about in his desk. "How about this one?"

" 'Give a problem where the answer is the same as the problem,'" the girl read.

"Take some time and—"

" 'Give a problem where the answer is the same as the problem,'" she said, thrusting the card back. "You'll have to do better than that."

The teacher sighed, and offered her another card.

" 'Call two associated natural numbers a pair...' Okay, this one looks a bit meatier."

"Something to tide you over during spring break."

"This'll do. Thanks! See you at the ceremony tomorrow!"

She flicked her fingers in a rapid Fibonacci sign, to which the teacher held up a hand, then waved as she headed out the door.

He leaned back in his chair and looked out the window. Any day now, it would be filled with cherry blossoms. Spring once again, the eternal cycle of the seasons. But he didn't feel trapped in a loop. He and his students were spreading their wings and flying upward. A spiral, then. Forever upward and onward, to destinations unseen.

What should be pointed out [to beginning students] is the truly astonishing number of simple and non-trivial theorems and relations that prevail in mathematics.... In my opinion, this property of mathematics somehow mirrors the order and regularity which appears in the whole world, which turns out to be incomparably greater than would appear to the superficial observer.

KURT GÖDEL
quoted in *Logical Dilemmas: The Life and Work of Kurt Gödel* by John Dawson

Afterword

"Every book has an intrinsic impossibility, which its
writer discovers as soon as his first excitement
dwindles. The problem is structural; it is insoluble; it
is why no one can ever write this book."

ANNIE DILLARD
The Writing Life

Thank you for reading *Math Girls 3: Gödel's Incompleteness Theorems*. This is the third volume in the *Math Girls* series, a sequel to *Math Girls* (2007) and *Math Girls 2: Fermat's Last Theorem* (2008). As with the previous books, this is a story in which Miruka, Tetra, Yuri, and the narrator explore mathematics and youth.

When I began this book I thought I understood Gödel's incompleteness theorems pretty well, but the more I wrote the more evident the sloppiness of my understanding became. So I hit the books, and after a year of careful study and guidance from many helpful supporters I think I finally get it. If you find any mathematical errors in this book, I would be very grateful if you would let me know.

As with the other books in the *Math Girls* series, this book was created using LaTeX 2_ε and the AMS Euler font. Also as before, Haruhiko Okumura's book *Introduction to Creating Beautiful Documents with* LaTeX 2_ε was an invaluable aid during layout, and I thank him for it. I created most of the diagrams using an elementary mathematics handout macro, emath, created by Kazuhiro

Okuma (a.k.a. tDB). Other diagrams were created using Microsoft Visio.

While I was writing this book, the publisher Media Factory released a comics edition of *Math Girls*. I thank illustrator Mika Hisaka and editor So Yurugi for helping to bring *Math Girls* to an even wider audience.

I would also like to thank the following persons for proofreading and giving me invaluable feedback. Of course, any remaining mathematical errors are solely the responsibility of the author.

> Hiroshi Fujita, Daiki Hagiwara, Hiroaki Hanada, Yoichi Hirai, Tatsuya Igarashi, Hiromichi Kagami, Toshiki Kawashima, Takayuki Kihara, Kayo Kotaki, Masahide Maehara, Tadanori Matsuki, Takashi Matsumoto, Kohei Matsuoka, Aya Miyake, Kiyoshi Miyake, Kenta Murata (mrkn), Masato Okada, Tamami Soma, Ryuhei Takata, Haruaki Tasaki, Ryuhei Uehara, Kenji Yamaguchi, Yuko Yoshida

I would like to thank my readers and the visitors to my website, and my friends for their constant prayers.

I would like to thank my editor, Kimio Nozawa, for his continuous support during the long process of creating this book.

I thank the many readers of the *Math Girls* series for their support. Your encouragement is a treasure beyond all others.

I thank my dearest wife and my two sons.

I dedicate this book to Kurt Gödel for the wonderful new paths he forged, and to all mathematicians everywhere.

Finally, thank *you* for reading my book. I hope that someday, somewhere, our paths shall cross again.

Hiroshi Yuki

2009

http://www.hyuki.com/girl/

Recommended Reading

> "Well of course I studied. I learned most of what I
> know from books."
>
> ――――――――――――――――――――――――――――――
> CEZAIMARU VENICO
> in *Six Supersonic Scientists*, Hiroshi Mori

This section is divided up as follows:

· General reading

· Recommended for high school students

· Recommended for college students

· Recommended for graduate students and beyond

· Web pages

These classifications are meant only as a guideline. Some texts may
be more or less challenging depending on the level of the reader.

> [Note: The following references include all items
> that were listed in the original Japanese version
> of *Math Girls 3: Gödel's Incompleteness The-
> orems*. Most of those references were to Japanese
> sources. Where an English version of a reference
> exists, it is included in the entry.]

GENERAL READING

[1]   Yuki, H. (2011). *Math Girls*. Bento Books. Published in Japan as *Sūgaku gāru* (Softbank Creative, 2007).

> The story of two girls and a boy who meet in high school and work together after school on mathematics unlike anything they find in class.

[2]   Yuki, H. (2012). *Math Girls 2: Fermat's Last Theorem*. Bento Books. Published in Japan as *Sūgaku gāru / Ferumā no saishū teiri* (Softbank Creative, 2008).

> The second book in the *Math Girls* series, where new "math girl" Yuri joins the others on a quest to understand "the true form" of the integers. Presents groups, rings, and fields among other topics, building to a tour of Fermat's last theorem.

[3]   Hofstadter, D. (1985). Gödel, Escher, Bach: An Eternal Golden Braid. Basic Books. Translated by Nozaki, A., et al. as *Gēderu, Esshā, Bahha—Arui wa fushigi na kan* (Hakuyo, 1985).

> A book about self-reference, recursion, representation of knowledge, artificial intelligence, and many other topics, taking Gödel, Escher, and Bach as its theme. This book was a reference throughout this book. The epigraph at the beginning of chapter 7 was inspired by this book.

[4]   Nozaki, A. and Anno, M. (1984). *Akai Bōshi* [The Red Hat]. Dowa-ya.

> A lovely picture book that I used as a reference for the hat-guessing game in chapter 1.

[5] Rényi, A. (1967). *Dialogues on Mathematics*. Holden Day. Translated by Yoshida, J. as *Sūgaku ni tsuite no mittsu no taiwa: Sūgaku no honshitsu to sono taiō* (Kodansha, currently out of print).

> A book that uses an imagined conversation between Socrates, Archimedes, and Galileo to explore fundamental questions such as what it is that mathematics studies, and how it benefits us.

[6] Lindbergh, A. (1955). *Gift from the Sea*. Pantheon Books. Translated by Yoshida, K. as *Umi kara no okurimono* (Shinchosha, 1967).

> A book about the importance of living simply, of valuing the time one has alone, and being fulfilled with a small number of things. Discussions of the sea as an allegorical reflection on daily life itself.

RECOMMENDED FOR HIGH SCHOOL STUDENTS

[7] Nozaki, A. (2006). *Fukanzen teiri: Sūgakuteki taikei no ayumi* [The Incompleteness Theorems: A Walk Through Mathematical Systems]. Chikuma Shobo.

> A book that starts from the early history of mathematics, explaining sets, logic, formal systems, and metamathematics to give a bird's-eye view of the incompleteness theorems.

[8] Shiga, K. (1994). *Kyokugen no fukami: Sūgaku ga sodatte iku monogatari 1* [The Depth of Limits: The Story that Math Tells, Vol. 1]. Iwanami Shoten.

> An easygoing book that uses alternating sections of text and mathematical notation to tell the story of mathematics. I used this book as a reference for chapters 4 and 6.

[9]   Tajima, I. (1978). *Epushiron deruta* [Epsilon–Delta]. Kyoritsu
      Shuppan.

      A text that focuses on the (ϵ, δ) definition
      of limits. I used this book as a reference for
      chapter 6.

[10] Kohari, A. (1996). *Sūgaku* I·II·III $\cdots \infty$: *Kōkō kara no sūgaku
      nyūmon* [Mathematics I · II · III $\cdots \infty$: A Mathematics Primer
      for High School and Beyond]. Hyōronsha.

      An exploration of mathematical problems
      through silly conversations between the
      book's characters. I used this book as a ref-
      erence for the hat problem in chapter 1.

[11] Takeuchi, G. (2001). *Shūgō to wa nanika: Hajimete manabu
      hito no tame ni* [What are Sets? A Book for New Learners].
      Kodansha.

      A well-regarded book that explains set the-
      ory. I used this book as a reference for chap-
      ter 3.

[12] Adachi, N. (2000). *Mugen no paradokkusu* [The Paradoxes of
      Infinity]. Kodansha.

      A book that gives broad explanations of in-
      finity. I used this book as a reference for chap-
      ter 4.

[13] Shiga, K. (2008). *Mugen e no hishō: Shūgōron no tanjō* [Fly-
      ing to Infinity: The Birth of Set Theory]. Kinokuniya Shoten.

      A book that explains set theory by tracing
      through Cantor's ideas. I used this book as a
      reference for chapter 3.

[14] Yoshida, T. (2000). *Kyosū no jōcho—Chūgakusei kara no
      zenpōi dokugaku-hō* [The Moods of Imaginary Numbers—All
      About Self-Study From Middle School and Beyond]. Tokai Uni-
      versity Press.

A massive volume on becoming self-motivated in "learning by doing" from the very basics, with a special emphasis on math and physics. A wonderfully interesting book. I used this book as a reference for the spiral graph in chapter 9.

[15] Yuki, H. (2005). *Purograma no sūgaku* [Math for Programmers]. Softbank Creative.

A book that teaches the mathematics needed for computer programming, avoiding the use of mathematical notation wherever possible. I used this book as a reference for chapter 7.

RECOMMENDED FOR COLLEGE STUDENTS

[16] Gödel, K. (1931). *On Formally Undecidable Propositions of* Principia Mathematica *and Related Systems*. Translated by Hayashi, S. and Yasugi, M. as *Fukanzen teiri*. (Iwanami Shoten, 2006).

The original paper by Gödel that this book is primarily concerned with. The Japanese translation includes commentary and a summary of the results of Hilbert's research. I used this as a reference throughout this book.

[17] Kato, K. (1985). *Gēderu no sekai: Kanzensei teiri to fukanzensei teiri* [The World of Gödel: His Completeness Theorem and Incompleteness Theorems]. Kaimeisha.

A book that contains Japanese translations of the papers in which Gödel presents his completeness theorem and his incompleteness theorems, along with easy-to-understand commentary. I used this as a reference throughout this book.

[18] Maehara, S. (2006). *Sūgaku kisoron nyūmon* [An Introduction to the Basic Theories of Mathematics]. Asakura Shoten.

> A reprint of a 1977 textbook that presents a careful introduction to the paper in which Gödel presents his incompleteness theorems.

[19] Matsumoto, K. (2001). *Fukkan sūri ronri gaku* [Mathematical Logic (Reprint)]. Kyoritsu Shuppan.

> A textbook for mathematical logic.

[20] Ishitani, S. (2006). (ϵ, δ) *ni naku* [Crying Over (ϵ, δ)]. Gendai Sūgakusha.

> A collection of easy-to-misunderstand topics in mathematics, including (ϵ, δ). I used this as a reference for chapter 6.

[21] Adachi, N. (2002). *Kazu: Taikei to rekishi* [Numbers: Systems and Their History]. Asakura Shoten.

> As the book states from the beginning, it "starts with logic, and ends with the introduction of complex numbers" to build a system of numbers from the ground up. A curious book, in that while its stated goal is simply developing a system of numbers, it repeatedly touches upon fundamental mathematical concepts while doing so.

[22] Shimauchi, T. (2008). *Sūgaku no kiso* [Fundamentals of Mathematics]. Nippon Hyōronsha.

> Similar to the above, as the book states from the beginning that it "starts with logic, and ends with elementary functions" to build a system of mathematics from the ground up.

[23] *Iwanami sūgaku nyūmon jiten* [The Iwanami Dictionary of Elementary Mathematics]. Iwanami Shoten.

A dictionary with easy-to-understand definitions of mathematical terms.

[24] Graham, R., Knuth, D., and Patashnik, O. (1994). *Concrete Mathematics: A Foundation for Computer Science (2nd Edition)*. Addison-Wesley Professional. Translated by Arisawa, M., Yasumura, M., Akino, T., and Ishihata, K. as *Konpyūtā no sūgaku* (Kyoritsu Shuppan, 1993).

A textbook on discrete mathematics, with
finding sums as its theme. I referenced this
book when writing chapters 8 and 9.

[25] Gries, D. and Schneider, F. (1993). *A Logical Approach to Discrete Math*. Springer. Translated by Shibagaki, W., Shimizu, K., and Tanaka, Y. as *Konpyūtā no tame no sūgaku: Ronriteki apurōchi* (Nippon Hyoronsha, 2001).

A book on discrete mathematics with the
goal of learning to use logic as a tool for
thought. It includes a huge number of practice problems. I used this book as a reference
when writing chapter 2.

[26] Aigner, M. and Ziegler, G. (2003). *Proofs from THE BOOK*. Springer. Translated by Kanie, Y. as *Tensho no shomei* (Springer-Fairlark Tokyo, 2002).

A collection of beautiful theorems and proofs
from various areas of mathematics. Begun
by Paul Erdős, famous for his collaborations
with mathematicians throughout the world,
the authors have added to his original collection.

[27] Takenouchi, O. (1971). *Nyūmon shūgō to isō* [An Introduction to Sets and Wave Phases]. Jikkyō Shuppan.

A book about set theory. I used this book as
a reference when writing about a function to

provide a correspondence between $0 < x < 1$ and the set of all real numbers, and the problem of the uncountability of real numbers in chapter 7.

RECOMMENDED FOR GRADUATE STUDENTS AND BEYOND

[28] Tanaka, K. (Ed.) (2006). *Gēderu to nijūseiki no rojikku 1: Gēderu no nijūseiki* [Gödel and Twentieth-century Logic, Vol. 1: Gödel in the Twentieth Century]. University of Tokyo Press.

Published in commemoration of Gödel's one-hundredth birthday, this series looks back on the development of logic in the twentieth century. Volume 1 gives an overview of the history of logic, and touches on the philosophical debate surrounding Gödel's incompleteness theorems.

[29] Tanaka, K. (Ed.) (2006). *Gēderu to nijūseiki no rojikku 2: Kanzensei teiri to moderu riron* [Gödel and Twentieth-century Logic, Vol. 2: The Completeness Theorem and Model Theory]. University of Tokyo Press.

Volume 2 in this series presents an explanation of Gödel's completeness theorem, model theory, and semantics.

[30] Tanaka, K. (Ed.) (2007). *Gēderu to nijūseiki no rojikku 3: Fukanzen teiri to sanjutsu no taikei* [Gödel and Twentieth-century Logic, Vol. 3: The Incompleteness Theorems and Systems of Arithmetic]. University of Tokyo Press.

Volume 3 in this series describes both the first and second incompleteness theorems. It also provides an explanation of reverse mathematics, which classifies theorems according to the axioms required to prove them.

[31] Tanaka, K. (Ed.) (2007). *Gēderu to nijūseiki no rojikku 4: Shūgōron to puratonizumu* [Gödel and Twentieth-century Logic, Vol. 4: Set Theory and Platonism]. University of Tokyo Press.

> Volume 4 in this series describes set theory and the mathematical philosophy of Gödel.

WEBSITES

[32] Yuki, H.: *Math Girls* http://www.hyuki.com/girl/en.html.

> The English version of the author's *Math Girls* web site.

[33] Kamo, H.: */dev/wd0a* http://d.hatena.ne.jp/wd0/.

> I referenced an article about the incompleteness theorems hosted at this web site.

[34] Hirzel, M.: *canon00-goedel.pdf* http://www.research.ibm.com/people/h/hirzel/papers/.

> An English translation of Gödel's paper. I referenced this paper because the functions and predicates are written in easy-to-read English.

> "You've gotta be brave enough to see what you aren't getting, and to take the time to think things through until you've mastered it. Once you've done that, it's yours forever. And that's a pretty nice feeling."

HIROSHI YUKI
Math Girls 3: Gödel's Incompleteness Theorems

Index

Other works by Hiroshi Yuki

(in English)

- *Math Girls*, Bento Books, 2011

- *Math Girls 2: Fermat's Last Theorem*, Bento Books, 2012

- *Math Girls Manga*, Bento Books, 2013

- *Math Girls Talk About Equations & Graphs*, Bento Books, 2014

- *Math Girls Talk About the Integers*, Bento Books, 2014

- *Math Girls Talk About Trigonometry*, Bento Books, 2014

(in Japanese)

- *The Essence of C Programming*, Softbank, 1993 (revised 1996)

- *C Programming Lessons, Introduction*, Softbank, 1994 (Second edition, 1998)

- *C Programming Lessons, Grammar*, Softbank, 1995

- *An Introduction to CGI with Perl, Basics*, Softbank Publishing, 1998

- *An Introduction to CGI with Perl, Applications*, Softbank Publishing, 1998

- *Java Programming Lessons (Vols. I & II)*, Softbank Publishing, 1999 (revised 2003)

- *Perl Programming Lessons, Basics*, Softbank Publishing, 2001

- *Learning Design Patterns with Java*, Softbank Publishing, 2001 (revised and expanded, 2004)

- *Learning Design Patterns with Java, Multithreading Edition*, Softbank Publishing, 2002

- *Hiroshi Yuki's Perl Quizzes*, Softbank Publishing, 2002

- *Introduction to Cryptography Technology*, Softbank Publishing, 2003

- *Hiroshi Yuki's Introduction to Wikis*, Impress, 2004

- *Math for Programmers*, Softbank Publishing, 2005

- *Java Programming Lessons, Revised and Expanded (Vols. I & II)*, Softbank Creative, 2005

- *Learning Design Patterns with Java, Multithreading Edition, Revised Second Edition*, Softbank Creative, 2006

- *Revised C Programming Lessons, Introduction*, Softbank Creative, 2006

- *Revised C Programming Lessons, Grammar*, Softbank Creative, 2006

- *Revised Perl Programming Lessons, Basics*, Softbank Creative, 2006

- *Introduction to Refactoring with Java*, Softbank Creative, 2007

- *Math Girls / Fermat's Last Theorem*, Softbank Creative, 2008

- *Revised Introduction to Cryptography Technology*, Softbank Creative, 2008

- *Math Girls Comic (Vols. I & II)*, Media Factory, 2009

- *Math Girls / Gödel's Incompleteness Theorems*, Softbank Creative, 2009

- *Math Girls / Randomized Algorithms*, Softbank Creative, 2011

- *Math Girls / Galois Theory*, Softbank Creative, 2012

- *Java Programming Lessons, Third Edition (Vols. I & II)*, Softbank Creative, 2012

- *Etiquette in Writing Mathematical Statements: Fundamentals*, Chikuma Shobo, 2013

- *Math Girls Secret Notebook / Equations & Graphs*, Softbank Creative, 2013

- *Math Girls Secret Notebook / Let's Play with the Integers*, Softbank Creative, 2013

- *The Birth of Math Girls,* Softbank Creative, 2013

- *Math Girls Secret Notebook / Round Trigonometric Functions*, Softbank Creative, 2014

- *Math Girls Secret Notebook / Plaza of Sequences*, Softbank Creative, 2014

CPSIA information can be obtained
at www.ICGtesting.com
Printed in the USA
FSHW010641040319
56076FS